Divided Loyalties

This case is a pioneer example of government steering not rowing and some of the problems this involves ethically + administratively.

Science and Society:
A Purdue University Series in Science, Technology, and Human Values

Leon E. Trachtman, General Editor

Volume 4

Divided Loyalties

Whistle-Blowing at BART

by Robert M. Anderson,
Robert Perrucci,
Dan E. Schendel, and
Leon E. Trachtman

Purdue University
West Lafayette, Indiana
1980

Any opinions, findings, conclusions, or recommendations expressed herein are those of the authors and do not necessarily reflect the views of the National Science Foundation.

Library of Congress Card Catalog Number 79-89588
International Standard Book Number 0-931682-09-6
Printed in the United States of America

Contents

Acknowledgments

The authors have a number of debts which they wish to acknowledge. We owe much to the National Science Foundation and to project monitor William Blanpied for financial support and moral encouragement of this project. Sandy Dukes offered a number of thoughtful opinions about the legal issues involved and also obtained all the relevant court documents. Jim Otten subjected the manuscript at several stages to the searching scrutiny of a professional philosopher and ethician. Lea Stewart, graduate research assistant who worked on the project from its inception, did a masterful job of developing filing systems for the huge body of materials which was assembled and drew up the detailed chronology and other documents upon which the authors were most dependent.

Marlene Mann and Donna Rosen were responsible for typing the manuscript into the computer and responded with alacrity and accuracy when asked to make the many revisions we required. Cathy Richardson typed many chapters of the manuscript and did numerous complicated and detailed searches for references. Laura Gitlin was most helpful in coding interview data, and Mary Perigo assisted in the preparation of early drafts of the manuscript. To all these, our heartfelt thanks and our hope that the final product justifies their efforts.

This volume was developed and produced with assistance of grants from the National Science Foundation Nos. OSS-76-14230 and OSS-79-20481.

Chapter 1

Organizations, Professions, and the Public Interest:

On the Social Control of Public Sector Organizations

Students of contemporary American Society who have sought to capture some of its central features have often described it as a "technological" or "organizational" society. Both terms reflect some of the more prominent changes that have occurred in American life in the last half-century. Large-scale organizations have attained a scope and centrality in our society on a scale never seen before. Ever-increasing areas of our lives take place within, or are controlled or affected by, large organizations. Economic life, health and welfare, education, and the security of our citizens and our society are dealt with in consciously created human inventions called organizations. When these organizations fail to achieve socially conceived and desired goals, the welfare of individuals and society suffers.

Paralleling the growth in the scope and centrality of large organizations, and indeed, one of the basic causes of such growth, have been rapid advances in technology related to the production and transportation of goods and to the generation and dissemination of information. Accompanying these advances in technology have been the appearance and rapid growth in significance of the science and technology-based professions. Members of the traditional professions of law, medicine, and the ministry have been joined by the scientists, engineers, computer analysts, and managers who make similar claims of "service to humankind" in return for which they expect to receive the honors, powers, and rewards that a society bestows on its professionals.

What is especially distinctive about members of these "new" professions is that they carry out their activities in large organizations, rather than as "solo" professionals, and they normally serve a collective rather than an individual client. The traditional professions deal with individual clients who are personally experiencing medical, emotional, spiritual, or legal crises. The professional-client relationship in these cases must be characterized by a high degree of trust on the client's part that the professional will use his knowledge and his power over the client to serve the clients' best interests. A misuse of that knowledge or power can seriously affect the client, but can have only minimal consequences for society as a whole. A client who is unhappy with a professional's advice or work can give his business to another professional, or in the extreme case, may even bring a lawsuit to correct any damage incurred in their relationship.

The new science and technology-based professionals, however, carry on their work as members of an organization, usually a very large complex organization such as a university, corporation, government agency, or nonprofit institute. Moreover, rather than serving the interests of any individual, their work is either designed specifically to serve the public welfare, for example, by constructing a mass transportation system, or it is designed to serve directly certain aggregates of private interests, stockholders, for example, while having indirect consequences for the public welfare, e.g., offering goods or services for public consumption, and possibly incurring social costs, e.g., creating air or water pollution.

Increasingly, the newer professionals are working in public and private organizations that are involved with such broad societal issues as planning of urban areas, controlling environmental pollution, improving public transportation, enhancing public education, and developing new communications systems. The decades ahead will find science and technology-based professionals working on such universal problems as developing large-scale systems in the production control, and use of energy and natural resources, planning new methods of food production and distribution, and developing biosocial systems, concerned not only with generating new concepts of medical care, housing and pollution control, but also with coordinating these advances into large-scale social systems, such as the modern megalopolis.

In addition to the cadre of professionals which has been developed for employment complex, private sector organizations, these future programs in system design will call for the involvement of increasing numbers of science and technology-based professionals in public sector organizations designed to serve the needs of the general public. Employment in the public sector will emphasize the ethical obligation of these professionals to serve the broad public interest as well as the more parochial interests of their respective organizations. It will also require that organizations be structured in a way that accepts professionals as contributing organization members while recognizing their responsibilities to the public-at-large as legitimate professional concerns. Moreover, since public sector organizations operate with goal structures, standards of evaluation, and exposure to public scrutiny different from most private sector organizations, such differences may require new organizational structures appropriate to them.

The general trends described above provide the context in which we can examine the profound and far-reaching questions of how public-sector organizations and their individual members can operate to achieve their corporate goals and still be held ethically accountable for their activities. Ultimately, this inquiry inevitably leads to basic questions about the control of public-sector organizations. Since the law has had only limited impact on the internal structure and activities of private corporations, can we assume similar limitations with respect to public or third sector organizations, or is there a body of law which can be used to regulate the structure and operation of these non-private organizations? Further, is there a public interest role that can be played by professional and technical societies in assisting public-sector organizations, and professionals in those organizations, when there is conflict between employees and employers? More basically, just what are the professional and ethical responsibilities of scientists, engineers, and managers in trying to fulfill their dual obligations to their organizations and to the public-at-large? What are meaningful standards of efficient operation for public organizations which commonly operate without the ultimate test of the market place that private sector organizations must face?

These questions facing professionals, professional societies, and public sector organizations are revealed and studied in all their complexity in this book through an examination of a particular organization and a particular incident, from which some generalizations or answers are derived. On March 2 and 3, 1972, three engineers were fired from the San Francisco Bay Area Rapid Transit District (referred to hereafter as BART). These engineers were reported to have had concerns that the automatic train control system of what was being hailed as America's first truly space-age mass transportation system was not safe enough. They repeatedly took their concerns to management, but were not satisfied with the response. Thereupon they carried the issue privately to certain members of the Board of Directors and to outside professional consultants, who made a public disclosure of the engineers' concerns. The engineers were fired. They had become "whistle-blowers", joining the growing ranks of professionals employed in private corporations and public agencies who have felt it necessary to expose what they believed to be illegal, inefficient, immoral or unethical practices of their employing organizations.

Simply to treat this case as another incident of "whistle-blowing" does not help us to understand either why such incidents are occurring with increasing frequency or what the basic organizational and professional structures are which tend to lead to such incidents. There is a widespread tendency to view "whistle-blowing" as idiosyncratic; as an aberration of an otherwise healthy, efficient organizational structure that typically achieve their corporate objectives while satisfying the expectations of its members or employees. In this book, we take a different view. We believe that the publicized instances of "whistle-blowing" are only the tip of the iceberg. Moreover, we believe that acts of "whistle-blowing" cannot be dismissed, or even understood, by simple personality analyses. "Whistle-blowers" are not disgruntled misanthropes bent on displacing their own career failures on their supervisors or their organizations; nor are they ethical paragons who are different from, and certainly better than, the other mortals who work in organizations and have compromised their principles and remained silent; nor do incidents of whistle-blowing occur only in repressive fascist-type organizations which force this deviant behavior on mistreated employees.

Rather, we take the view that an act of "whistle-blowing" is the culmination of a series of actions and reactions that take place in the course of a professional's relationship with an employing organization. Therefore to understand "whistle-blowing" it is necessary to identify the organizational conditions--authority structure, lines of communication, opportunities to participate in decision-making--that give rise to initial acts of disagreement with, or concern about, some organizational practice. Further, to understand whistle-blowing adequately, we feel it necessary to trace the sequence of events that lead a professional to follow one course of action over others and that lead superiors to respond to those actions in one way rather than another. Finally, we feel it is essential to place the principal actors in a "whistle-blowing" incident (i.e. a professional and his/her superiors) in a larger context of economic, political, and professional forces impinging upon the situation. Managers in a corporate structure experience points of tension with Directors over issues of control and authority. Directors operate in a political context of forces external to the corporation seeking to influence or shape its operations. Professional employees work in a setting with other colleagues and as members of professional associations, both of which influence their professional self-concepts and their views about proper conduct as employees and professionals. And professional associations are composed of interest groups with varying views about the proper role of professionals as employees, and about the responsibilities of professional societies to their members.

The organization studied in this book, like other private and public sector organizations, develops strategies and structures designed to reduce uncertainty in its internal environment, and attempts to establish favorable power relationships with its external environment in order to protect autonomy. Figure 1-1 summarizes the different elements and forces found in any organization's internal and external environment.

Forces in the external environment impinge upon and attempt to constrain the actions of the organizations. Sometimes these forces attempt to shape the organization as a totality by setting conditions under which it can and cannot operate, e.g., environmental protection legislation or federal regulatory law. At other times, external forces are mediated through specific groups in the internal environment, as in the case of professional societies, unions,

FIGURE 1-1
INTERNAL AND EXTERNAL ENVIRONMENTS
OF ORGANIZATIONS

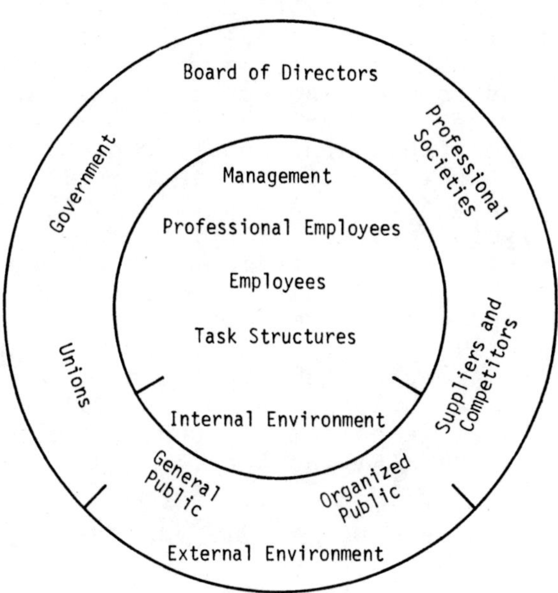

or special interest groups. Most organizations contain specific positions in their internal environment whose sole responsibility is to interact and relate with those sectors of the external environment that are most important to the organization.

At times, the transactions, interactions, and political negotiations between the internal and external environment are so important and continuous that the attempt to distinguish between what is or is not part of the internal environment of an organization is meaningless.

Our approach to the analysis of a specific instance of whistle-blowing in a public sector organization provides an opportunity for intensive analysis of the interplay of power relationships within organizations and between organizations and their environments. Our analysis has been informed by several theoretical perspectives on organizations and by methodological approaches to implement those perspectives. A brief discussion of these perspectives will be helpful in understanding our analysis of the BART incident.

Organizations as Private Governments

Among some of the more entertaining anecdotes about the modern corporation are those testifying to the truly incredible scale of activities in which they are engaged. We are told, for example, that the total assets of General Motors exceeds the gross national product of many of the member states of the United Nations. The full significance of this rather common economic fact can be grasped only if you are prepared to view General Motors as something more than a business firm. Many corporations also provide a full range of governmental services for the employee and his/her family, such as health and retirement plans, family education programs, recreational facilities, and retirement villages. In short, the organization may come close to providing a total life environment and experience for its employees.

Large organizations can be viewed, in many respects, as private governments. They are concerned about their autonomy and their sovereignty in relation to such other organizations in their environment as labor unions, government, professional associations, and other private governments (i.e. corporations). Negotiating with elements of their environment is analogous to having a "foreign policy" and "international relations." Similarly, the internal life of large organizations is structured by a particular political system that defines legitimate authority, the rights and responsibilities of its "citizens," the processes for dealing with conflicts, the rewards for good "citizens," the punishments for bad "citizens," and the scope of influence the "state" has over its "citizens." Some organizations approximate totalitarian states, and others approximate democracies, and others are loose federations with regard to the rights of "citizens" and the limitations on the "state's" intrusion in their lives.

While we do not wish to carry the nation-state metaphor too far, we do wish to convey our belief in the usefulness of viewing organizations as political-economic systems, containing political factions and interest groups, governed by political processes, and faced from time to time by political activists, revolutions, and coups d'etat.

An examination of the question of the social control of the modern corporation is provided by Stone (1975) who assesses the effectiveness of the market, the law, and the board of directors in controlling the actions of corporations. Social control processes and the question of employee rights within organizations have been examined in a recent work by Ewing (1977a) who points to the abridgment of basic rights in many corporate actions. However, the approach to organizations from the perspective of the nation-state analogy is best developed in the writings of Weinstein (1977), and Zald and Berger (1978) who urge an analysis of bureaucratic opposition and conflict in terms of conventional and unconventional political relationships between ruling powers and organized internal insurgencies and social movements seeking change.

Thus, in our study, we approach the Bay Area Rapid Transit System as something other than a rational, decision-making structure which attempts to attain its goals through the application and coordination of positions filled by persons with expert knowledge. As an organization, BART emerges out of a political struggle and it tries to sustain itself via political processes.

Organization Structure, Authority, and Professional Employees

The growth of the science and technology-based professions and the increasing employment of scientists and technologists in large-scale industrial and governmental organizations have led many scholars to identify and study some persistent sources of tension and conflict between organizational needs and professional requirements (Marcson, 1960; Kornhauser, 1963; Glaser, 1964; Scott, 1966; Perrucci and Gerstl, 1968). Organizations are described by these scholars in terms of their need for predictable behavior, coordination of activities, loyalty to the organization, and the use of economic criteria in selecting organizational goals. Professionals, on the other hand, are said to emphasize authority based upon knowledge rather than position, freedom from external control in matters related to their work, commitment to colleagues as the main source of professional rewards, and a strong service ethic or sense of social responsibility.

While this description of the central defining characteristics of organizations and of professional employees is overdrawn in its emphasis on points of tension rather than points of mutual accommodation and common goals, it does nonetheless identify a possible source of problems for organizations with professional employees. Thus, in the present study of the BART organization we shall be sensitive to the following questions involving the organization's structure and authority as they relate to its professional employees.

(1) Does the scope of organizational responsibility of professional employees influence their general level of job-career satisfaction? Organizations that emphasize a narrow specialization and fragmentation of tasks can experience considerable underutilization of professional employees, which is in turn related to job dissatisfaction (Ritti, 1971). Professional employees experiencing professional frustration because their skills are underutilized may be the most likely candidates to undertake criticism of their organization's practices and to oppose organizational authority.

(2) Are professional employees who have greater expert authority than positional authority more likely to resist, question, or oppose the organization? The structure of authority in an organization may reflect a wide range of consistency or inconsistency between positional and expert authority. Some organizations have persons in management positions who do not possess the technical-scientific competence of the professionals they supervise and direct. This sort of positional authority, unsupported by expertise, may be unstable and may be resisted by professionals who will view themselves as being subjected to "illegitimate" authority.

Is it possible that BART's problems arose because the organization had "built-in" authority problems based upon its choice of personnel in supervisory and middle-management positions?

(3) To what extent are professional employees brought into the decision-making process and kept fully informed of the broad range of economic and political issues that impinge upon their technical roles? To what extent should they be? In organizations such as BART, the goal being pursued is defined as non-routine and very exceptional. Professionals

recruited to organizations pursuing such exceptional goals are often asked to exhibit exceptional dedication, usually in the form of a strong personal commitment to the organization and its goals. In such a case, it may be essential that the involved professionals be allowed to participate to the fullest extent in a wider array of discussions and decisions related to organization goals than would be allowed in a more routine position. Can this participation take place in the context of the classically structured organization?

(4) Are the channels of communication and flow of information up and down the organizational hierarchy clearly understood and open in both directions? In organizations dealing with routine tasks and a well-known technology, the flow of information is generally from the top down, especially with regard to task-related decisions. However, in an organization like BART it may be more necessary to have information flow from the bottom up and be as unrestricted as that coming from the top down.

Professionals who identify with the goals of their organizations, especially when they are related as exceptional "breakthrough" type projects (e.g. space-age mass transportation system), may expect that their ideas, concerns, and suggestions will reach and be considered by the highest technical and administrative positions in the organization. They may also expect that any rejection of their ideas will be accompanied by a full technical and organizational explanation.

(5) To what extent did the organization structure built by the BART Directors and Management provide an effective fit with the goal being pursued? Such a question can, of course, be asked of any organization. What makes it especially important in this case is that BART was more than just an organization, it is an interorganizational network of organizations. During the developmental phase, this network was composed of the BART organization with its directors, management and professional

staff; the purposely created consulting firm of Parsons-Brinckerhoff-Tudor-Bechtel which was responsible for putting the system in place; the engineering firms which obtained the contracts for specific parts of the system (e.g. Westinghouse, the contractor for the automatic train control system); and the California State legislature, which was involved in providing the public funds for BART. Figure 1-2 provides a general description of this interorganizational network.

Complex networks require explicit coordination processes and periodic monitoring and evaluation of the performance of each component. In the BART case, we have what appears to be partial network in that all of the constituent organizations were not linked by a primary "tie" but instead had to work through another organization. Thus, the Board of Directors operated without an independent staff

FIGURE 1-2
BART'S INTERORGANIZATIONAL NETWORK

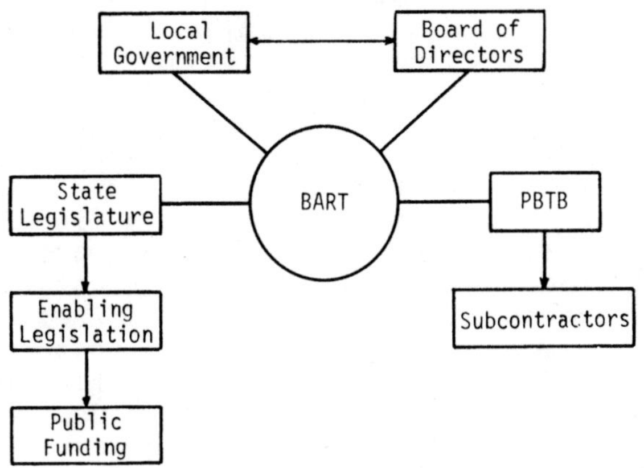

and line of communication that would permit them to monitor and assess independently the performance of PBTB or the subcontractors. Similarly, BART's knowledge of the performance of the subcontractors was filtered through PBTB, which clearly had an interest in the kind of information that was transmitted.

In our analysis of the BART case we will be concerned with examining how the influence of enabling legislation, and how the decisions made by management concerning development, coordination, and monitoring of this interorganizational network helped to lay the groundwork for many of the problems that eventually developed.

Professional Ethics and Professional Societies

In many of the cases involving professional employees' opposition to their organization, these employees believe that the organization is producing a product that is unsafe, that the organization is not complying with federal regulations, or that the organization is misusing its resources. In each instance, the professional is faced with a situation that he believes to be unethical. Feeling that he must take some sort of action to deal with an unethical situation the employee often looks to his professional society for guidance as to how to behave, or for protection if he fears that the organization may take action against him.

In our examination of the BART incident we shall be concerned with what was perceived by three engineers as an unethical situation which demand extraordinary action. Did their action follow inevitably from a specific code of ethics governing the behavior of a particular professional association, or is the code so ambiguous that several courses of action were possible and permissible? Can a professional believe he has satisfied his ethical responsibilities by informing superiors of the consequences of some practice or must actions go beyond this to seek corrective change in organization practices? Ethical codes specify the rights of professionals to protect the public interest by informing the "proper authority" when they believe there is danger to the public health or safety. But where is the proper authority? Is it the professional's immediate supervisor? Top management? The Board of Directors? Since the BART case involves a public-sector organization, is the proper authority the state legislature or the public-at-large? The growth of public sector organization also requires us to give greater attention to the question of how the public shareholder exercises ownership rights, legally and practically.

We shall also examine the structure of professional societies and how that structure may either facilitate or hamper their involvement in conflicts between professionals and employing organizations. This issue is confused by the built-in conflicts of interest of many professional societies, whose membership is often composed of a heterogeneous mix of practicing professionals, professionals who have moved into management positions, and owners of private consulting firms.

The issues regarding a professional's ethical obligation to protect the public welfare, and an employer's recognition and acceptance of this obligation, touch a much larger number of professionals and organizations than ever reaches public attention. In 1972, for example, the National Society of Professional Engineers surveyed a sample of its membership on the subject of ethical dilemmas (Olson, 1972). Over 10% of NSPE members responding answered "yes" when asked "if they were required to do things which violated their sense of right and wrong or if they felt that their employers interfered with their personal rights." Seven percent indicated that they had sought transfers or resigned positions "when asked to work on a product or project they believed not to be in the public interest," and 20% had refused projects or positions for the same reason. And finally, 40% reporting feeling "restrained from criticizing their employers' activities or products."

A survey of corporate leaders about their employees' rights by Ewing (1977) shows that, while top executives support greater protections for their employees who speak out on controversial issues, they also fear the long run costs associated with expansion of employee rights. Ewing reports that top executives "in the long run...worry more about threats to the social and economic order than about the suppression of individualism" (p. 82).

Thus, the issue of employee rights, employer recognition of those rights, and adequate protection for employees who take public stands on controversial matters are topics of immediate concern of professionals, corporations, professional societies, and legal scholars. We believe that our analysis of the BART incident will reveal the manner in which these issues are framed and disputed in reality.

We can now try to summarize what we hope to learn from this analysis of an incident of whistle-blowing in a single organization, and why we think such a undertaking is important.

(1) Whistle-blowing, as a recent publicized organizational phenomenon, is widely discussed but poorly understood. Most of what is available in published form consists of brief descriptive accounts of specific incidents that are either journalistic or highly partisan in form. We believe that our study is the only in-depth descriptive and analytical study of whistle-blowing available in scholarly literature that is based upon primary data obtained from a wide array of persons involved in the incident. A study such as the present one can facilitate examination of a broader class of incidents of whistle-blowing, as well as a comparison of organizational settings that seem to be related to the presence or absence of such incidents.

(2) The incident of whistle-blowing examined in this study provides an unexpected op-

portunity to see aspects of an organization's workings that are not revealed by study of its normal, recurring structures and activities. The incident of whistle-blowing serves as a critical incident, or crisis, in the life of an organization which illuminates aspects of organizational life often unrecognized under normal conditions.

We believe that this study reveals some of the issues and questions concerning the external political control of public sector organizations and the way in which that control often functions to create inefficiencies and ineffectiveness in goal-directed activities. Between enabling legislation, Boards of Directors, public shareholders, state legislatures, and organized public interest groups, there are so many competitors for control of the organization that it is often difficult to know "who is in charge." The lack of clarity of external centers of power and influence produces ambiguity with regard to power within the organization. This encourages the formation of competing coalitions between elements within the organization and those outside seeking greater control, which only contributes to the problems of public sector organizations.

(3) A third element in the present study is consideration of the importance of ethical problems faced by employed professionals, generally, and by those in public-sector organizations especially. The ambiguity concerning who is the "proper authority" in a public sector organization--management, the legislature, the public--can leave the concerned professional in an unacceptably unclear position when dealing with questions of professional ethics.

(4) The internal political life of an organization is the final element that is of central importance in the present study. The decision to oppose organizational authority, as in the case of whistle-blowing, is a political act designed to achieve a rearrangement of power relationships, leadership, or organizational policies. The recognition and analysis of such political phenomena contributes to an

understanding of organizational decision-making that takes us beyond rational, bureaucratic models.

Method: Actors' Viewpoints and Multiple Realities

Most readers are probably familiar with the Japanese film Rashomon , or with the particular idea upon which the film is based (see e.g., Goldstein, 1977). The film concerns four characters and an incident in which they are mutually involved. Each character in the film puts forward his view of what has taken place, obviously coloring the event in a way to enhance his own self-image. Yet each view is sufficiently clear, logical, and without internal contradiction that the viewer of the film is hard pressed to come to any firm conclusion about what actually happened. It is possible that each of the characters has so falsified what happened that there is no truth presented in the film. However, it is also possible that each character is telling his or her view of reality, and that the contradictions and differences between the various items are really the only truth that can be uncovered from the incident.

Early in the formative stages of this study of the BART incident we abandoned the idea of trying to come up with a single interpretation or explanation of the events surrounding the incident. Instead, we chose to allow each group of actors--the engineers, management, the directors and the professional societies--to present their independent viewpoints about the BART organization and the course of the incident of "whistle-blowing" with which we were concerned. In addition to reporting on their actions in this incident they were also asked to characterize the actions of others who were involved in it.

As a methodological strategy we have organized this book around the viewpoints of individual and collective "actors." In so doing, we have consciously chosen to play down the search for who or what was responsible for the incident, and have chosen instead to try to understand more profoundly just what took place and why it happened as it did. To understand a particular social phenomenon requires knowledge of the subjective understandings of the actors involved in the situation, and insight into how these subjective states are a response to their perceptions of the actions of other actors.

The plan of the book also reflects the methodological strategy selected by the authors. Each part of the book deals with a particular time period in the "life" of the BART organization. Part I starts with the initial idea of creating a rapid mass transit system in the Bay Area and ends at the point that the three engineers, Holger Hjortsvang, Robert Bruder, and Max Blankenzee have joined the BART organization and perceived certain problems. In Part I each of the groups of actors presents their experiences with the organization during its formative period. Theoretically, we view these experiences as "preconditions" for the critical incident of "whistle-blowing" that is to occur. To view each chapter as setting our preconditions is, of course, to begin to think causally in that the particular choices made about how to build the BART organization, the political battles that had to be fought to create BART, and the personal biographies of the three engineers serve to shape subsequent events and to set the course of the "whistle-blowing" incident.

The second part of the book covers a relatively brief period of time--about ten months. During this period events move very rapidly and represent a series of interrelated actions and reactions by the various actors in the incident. The viewpoints presented in each of the four chapters in Part II allows us to develop an understanding of how the "whistle-blowing" action was a product of the complex sets of perceptions held by persons and groups in BART about each others actions and motives.

The materials contained in Part II should provide the reader with a good understanding of the political "underlife" of a large organization. If any further evidence is required, it should also dispel any beliefs held by students of organization that one can learn much about what goes on in an organization from examining its formal structure. Day to day life in organizations does not seem to be much influenced by the formal codes, rules, regulations, and rationally conceived task structures that can be found in organization charts and annual reports.

While the emphasis of Part II is on the internal politics of the BART organization, Part III shifts attention to political processes between BART and organizations in its environment. The chapters in this part of the book deal with the aftermath of the firing of the three engineers. During this time, the BART organization, its Directors and management, come under scrutiny and criticism from the California State legislature, professional societies, and newspapers.

Materials in these chapters tell us something about the politics of survival, as the BART organization attempts to maintain its autonomy in the face of attempts by external bodies to restrict that autonomy and increase the accountability of BART, as a public-service organization, to legislated and self-appointed bodies acting for the public interest.

A final note on method concerns the way the four authors "cut up" the incident for analyses. Each author was responsible for representing the actions and viewpoints of one of the groups of actors in the BART incident. This permitted each author to learn as much as possible about how each group experienced life in BART through repeated interviews with persons in each group and through immersion in the documents related to the role of each group in the BART incident. Thus, the author responsible for representing the engineers' actions and viewpoints did not interview any of the Directors or managers. There was, however, extensive discussion among the authors as each developed an understanding of the role played by his group in BART. Project meetings often took on the character of role playing as the views of different persons were represented by their advocates.

We believe that this method allowed us to develop a better understanding of the incident in question and to approach the "truth" of what happened as much as may be possible. That "truth" emerges from the viewpoints of each group as well as the independent standpoints of the authors that are interjected in each chapter as commentaries and analysis on events reported and interpreted by each "actor" in BART. Thus, the authors ideas did not become fused with those of the group we represented. In short, we did not "go native" entirely by "becoming" the persons we were representing. A check on how successful the authors were in maintaining such "detached involvement" is provided by the fact that each actor read the final manuscript of the study and was asked to comment on its validity.

Finally, our method was not used to avoid taking a position on the "truth"; rather, it permitted the most complete reconstructed exposition of events and interpretation of events by each actor.

Part I
Preconditions

Chapter 2

Creating the Organization:

The Directors' View

The decline during the 20th Century of the family-owned or entrepreneurial business has sparked intensive discussions in public, academic and government circles on the question of who controls or should control the modern corporation. Claimants to control of the private organization include stockholders, the Board of Directors, the appointed managers, financial institutions that extend loans to corporations, and the consuming public, through a variety of privately organized and government-supported public interest organizations. For public or third sector (quasi-public/quasi-private) organizations, the question of control may become even more complicated. The goal of these organizations is not fundamentally to make profit and hence, this relatively simple criterion of accountability is not available. Boards of Directors and managers of these organizations must theoretically be accountable to the public which supports them through direct taxes or tax-supported bond issues, but accountability in these terms often means no more than periodic review by voters for elective boards or by various political agencies for appointed boards.

The basic issue in principle is complicated in practice by wide variations in powers granted to Boards of Directors and managerial groups by enabling charters and legislation. The different kinds of allocation of authority and responsibility by these documents can have serious implications for the behavior of individuals functioning in various roles and at various levels of the organization.

In the decade following the end of World War II, San Francisco experienced the transportation traumas and problems common to most crowded, auto-choked American metropolitan areas. It was largely the problems of moving people and goods into and out of the city which caused San Francisco to become increasingly concerned during this period with its civic vigor and health.

To many residents of the Bay Area, the warm, golden hills in the East and North Bay Areas proved more appealing than San Francisco as a place to live. And if they were better to live in, why not, one day, better to work in as well? If enough individuals and businesses followed this line of thought, the decay of the urban center of the Bay Area was inevitable. This concern led many civic leaders to study the future transportation requirements in the Bay Area, and especially those of San Francisco itself. If movement into and out of the city could be made swift, safe and pleasant--if commuters could avoid the delays and frustrations of the clogged Bay Area highways--there was perhaps hope for saving the central city as the commercial, cultural and entertainment center of the Area. This belief led to the development of political pressure by local interest groups, and the pressure led to the creation by the California State Legislature of the San Francisco Bay Area Rapid Transit Commission, a body charged with studying the problem and recommending solutions to the Legislature.

After six years of study, the 26-member Commission endorsed and elaborated upon the 1947 recommendation of a joint Army-Navy Review Board that a transbay tube be constructed for the exclusive use of a high speed train system. Acting on the recommendations of the Commission, the California State Legislature in 1957 created the Bay Area Rapid Transit District (commonly called BART). BART was charged with designing, building, and operating a rail-based rapid transit system linking nine counties: Marin to the North, San Francisco and San Mateo on the peninsula, Alameda and Contra Costa to the East and Sonoma, Napa, Solano, and Santa Clara which formed an outlying belt circling the Bay Area on the North, East and South. (See Figure 1 for a map of the Bay Area). It was soon determined, however, that the four outlying counties could not be economically included in a master public transportation plan, and so none of them was active in any aspect of BART, although provision was made to annex territory from them if necessary. From the five counties which remained active in BART, an initial board of directors was appointed to represent them and the cities of the District.

Structure and Role of the Board

BART's first Board, as spelled out by the enabling legislation, was composed of 16 directors, four from Alameda County, four from San Francisco, three from Contra Costa, three from San Mateo and two from Marin County. The differences in representation simply reflected differences in population of the counties. The cities within each county had representatives appointed through a "city selection committee," which, in practice, was made up of the cities' mayors. The Board of Supervisors in each county appointed the remaining members. Of the original Board of 16, seven members were appointed by city selection committees and nine by Boards of Supervisors. The term of directors' appointments was four years, except for the original board, whose members' terms were staggered across two, three, and four year terms.

The Board of Directors, partly because of the series of events which this book relates, became an elective body in late 1974. Only one of the original 16 directors, George Silliman of Alameda County, served over the full 17-year period that the Board was appointive. Silliman was also a candidate for election in 1974. Indicative, perhaps, of public assessment of Board performance over the previous 17 years, is the fact that Silliman was not elected.

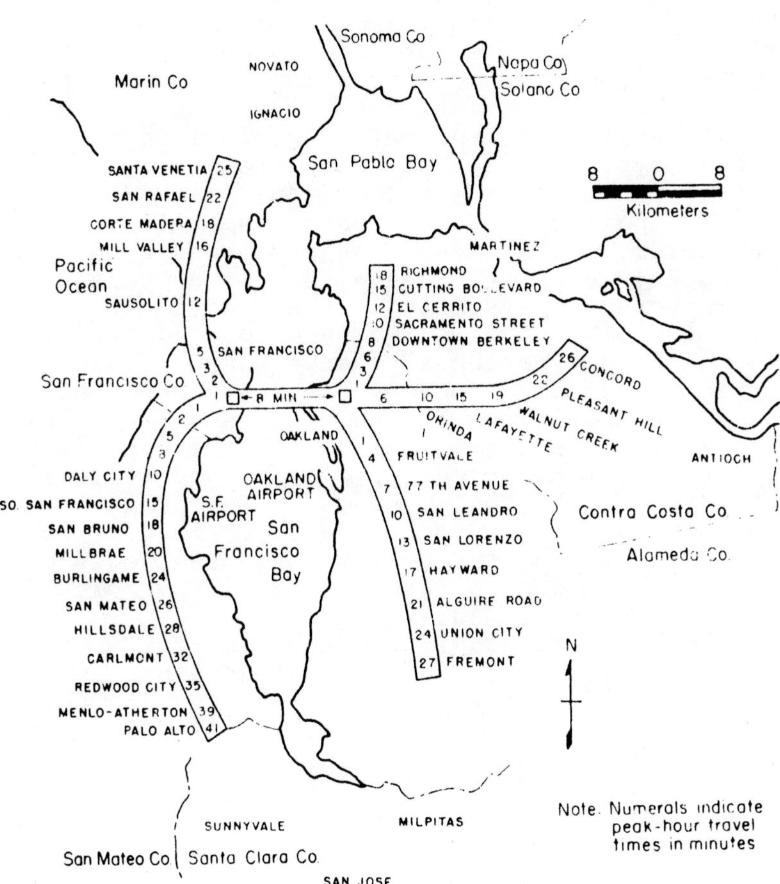

Figure 1: Five County Plan
For Bay Area Rapid Transit

The first major tasks of the Board were to place before the local electorate a design for BART and a proposal for bonding support for the system's construction, and to do these within a five year period, before the end of 1962. Both in BART's enabling legislation and in the completion of this first phase task can be seen the genesis of the later problems in professional ethics which this book examines.

The enabling legislation insured that BART's policy making Board was political in character. Each member was appointed by either a County Board of Supervisors or a Conference of Mayors representing the cities. Not surprisingly, therefore, each Board member was forced into a parochial stance when he had to vote on issues concerning the governmental or geographic unit which he represented. This political feature, coupled with the dependence of BART upon these same local governmental units for financing, forced the Board, at least on some policy issues, to face real difficulties in gaining the policy consensus necessary if the Board was to take a leadership role in running the railroad.

Next, although it appeared initially that BART's legislated fund raising authority was adequate to the task, in the long run it proved to be far too limited. The enabling legislation permitted BART to initiate bonding authority up to fifteen percent of the assessed valuation of taxable property within the District, and to tax the District to cover interest, principal, and sinking fund requirements for these bonds. Further, the BART Board could impose a property tax (Section 29123) "not to exceed five cents (.05) on each one hundred dollars ($100) of assessed valuation of taxable property within the District," and to levy use taxes. The Board was, however, restricted from raising more than 150 million dollars from any such taxes imposed. When the actual costs of completing the system turned out to be double the initial estimates these fund-raising provisions proved to be simply inadequate. It must be emphasized that it was not any niggardliness in the initial enabling legislation which caused BART's later financial problems. Rather, it was the unanticipated time delays and galloping inflation which doubled the expected cost of the system. For additional funds, BART had to go either to its local government parents and citizens (the cities and counties), the state legislature, or beyond that to the Federal level, despite the full intent of the Legislature to make this a local project, with local funding. BART's enabling legislation, therefore, while adequate and even liberal in intent, finally made a beggar out of BART, a fact that shaped its very design and future.

Third, the imagination of the populace had to be satisfied if a broad and stable pro-BART consensus was to be achieved. In an area that prides itself on being cosmopolitan, liberal in outlook, and modern in concept, this meant an aesthetically exciting and satisfying design which would also be forward looking and innovative in operation. George Silliman, a long-term board member, observed:

> We had to come up with a politically acceptable program. We had to come up with a design system acceptable to people and our biggest problem actually was coming up with a system that was advanced enough to be pleasing to people. In other words, when we talked about transit, for instance, we were fighting what was happening in New York, Philadelphia...the dirty, noisy, bad system, the problems with muggings. We were in a space age, so we had to come up with something that was really a space age type system.
>
> And I think this innovative system was one of the things we had problems with. The area once had a conventional surface transit system and they didn't want another one.

In addition to the key decision to develop a technically innovative transit system, the BART Board, along with its management team, made another significant policy decision when it elected not to hire and develop its own cadre of technical and support staff to design and build the system. Instead, in 1959 it formed a consortium of three separate engineering firms, initially to design and to negotiate the contracts for the system. Parsons, Brinckerhoff, Tudor and Bechtel (PBTB) was an amalgam of Parsons, Brinckerhoff, Hall and MacDonald, the New York-based transit consultants who did much of the work with the original 26-member transit commission, Tudor Engineering of San Francisco, and Bechtel, another San Francisco-based construction giant with an international reputation. Following the same line of reasoning, in 1962 BART's Board, rather than building an organization to serve as its own general contractor, selected PBTB for this task. Using a contract design team rather than employing one made a great deal of sense to the Board and, even in retrospect, is logical and understandable. The Board did not see how it could hire and build a technical team capable of developing the required innovative transit technology within the time and budget contraints it faced. After all, it would be many years before a BART organization would have operational responsibilities and the talent required to design and build the railroad appeared to be much different from that required to operate it.

Unfortunately, also in retrospect, it is possible to see that the decision to use a design engineering consortium might have helped create the basis for later problems in cost and time overruns, and in the safety issue related to the firing of the three engineers. What was not anticipated by the policy decision to use the consortium were the difficulties involved in making the transition from design, construction and testing of an innovative system to its routine operation. In BART's case, this technological transition required a transition in authority from the consultants and general contractor to the BART internal staff. Although the BART directors and general management thought they were buying a finished transit package which would require relatively simple operating and maintenance instructions to be relayed to BART's permanent staff, the transition turned out, in fact, to be a far more complicated matter.

Problems Arise

The BART concept suffered its first setback even before the voters of the area had a chance to pass on the system's bond issue. In 1962, the District shrank from five to three counties. First San Mateo County withdrew from the District, ostensibly because it already had the Southern Pacific serving its commuter demand. Marin County followed suit, on the basis of concerns about the stress which would be placed on the Golden Gate Bridge by BART trains. But both counties, it appears, were really more concerned with the pressure on their tax base than with technical issues. The withdrawal of these two counties left a three county district, and a dominance of East Bay counties in the political coalitions necessary to gain a majority on the Board. As a consequence, there was some hasty scrambling to redesign the BART concept which was to go before the voters for bond approval.

This bond issue was designed to finance the construction of the system adequately enough to secure its financial future. PBTB estimated a cost of $792 million, not including the Transbay Tube, with $186 million as a contingency reserve. The Transbay Tube cost was estimated at an additional $138 million. A key figure in the later events, B. R. Stokes, probably earned his right to become the General Manager of BART in 1963 through his work on publicizing and helping secure passage of this initial bond issue.

In California, a two-thirds majority is required to pass public bond issues. The groups pushing for the development of the BART system, aware of the substantial opposition to the idea, were hard at work in Sacramento to reduce the majority required for bond passage from two-thirds to sixty percent. Senator MacAtier from San Francisco was successful in placing a rider to this effect on another bill and, on June 6, 1961, the desired reduction became law.

Silliman recalls that the opponents of BART didn't pay much attention to the reduction in the majority required for passage because they believed it "didn't make any difference whether it was 60 or 66 percent; (BART) couldn't pass the bond issue anyway." As it turned out, the opponents were wrong. The issue passed--barely (San Francisco County passed by about 68 percent, Alameda about 61 percent and Contra Costa 54 percent, for a cumulative total of 61.2 percent for the three-county area.). The opponents of BART, it turns out, had underestimated by a razor-thin margin the appeal of the system. With this slight miscalculation, BART's opponents lost their last chance to stop the system.

In 1963, partially for his work on the bond issue, Bill Stokes became the General Manager of BART. In a sense, it took real boldness on the part of the Board to name a non-technical person to head this complex and technologically sophisticated enterprise. Stokes, a naval officer in World War II and for 12 years a reporter and writer for the _Oakland Tribune_, headed a BART staff which then numbered only some 25 people. But BART was finally off the ground.

Stokes, whose style, temperament, and drive were considered important to the further development of BART, began to take charge. Even though long time directors George Silliman and Harry Lange believe that, if they were to do it again, they probably would choose a technical man to run BART, neither would take anything away from Stokes. Of Stokes, Silliman says:

> I don't think we would have gotten as far as we did without Stokes. He had the drive, the initiative and the brass to go through with this

> thing. I think we would have folded with anybody else but Stokes. He could take adversity and stand there and let it roll off his shoulders.

For a dozen years Stokes led BART and, according to some, led the Board of Directors as well. He provided the leadership that created the BART organization and he instituted the contracting and consulting arrangements necessary to design, build, and finally operate the system. Although he left BART in 1974, before the first train went under the Bay to San Francisco, Stokes, more than any other group or individual, including the Board, created the organizational climate in which all of the events which we will examine took place.

The Climate of the Organization

What kind of a climate was it? The enabling legislation gave the General Manager these general powers: "(He) shall have full charge of the acquisition, construction, maintenance, and operation of the facilities of the district and also of the administration of the business affairs of the district." (Section 28830)

The General Manager was appointed indefinitely and did not perform under a contract. It took a majority vote of the Board to remove the General Manager, but even then, the Board had to hold a public hearing to show cause if the General Manager demanded it. These provisions gave the General Manager broad powers and great job security--perhaps, as some Board members could argue, necessary for such an enterprise as BART.

The Board, for its part, held the power and responsibility for all basic policy decisions. In terms of operations, the Board was responsible for approving all expenditures and was authorized to adopt a "personnel system" if it chose. Specifically, the enabling legislation stated:

> The Board may adopt a personnel system for the purpose of recruiting and maintaining an effective working force with good morale. The Board shall by resolution determine and create such number and character of positions as are necessary properly to carry on the functions of the district and shall establish an appropriate salary, salary range, or wage for each position so created. The Board may by resolution abolish any such position. Except as otherwise provided appointments to such positions shall be made by the General Manager. (Section 28767)

The General Manager, already given broad discretion in operating BART, had his power, freedom and discretion enhanced by the rules governing Board meetings. Simply put, any confrontation between the Board and the General Manager put the Board at a disadvantage in that it had to operate publicly. This disadvantage was compounded by Stokes, who used the power given him aggressively, and by a Board which was generally compliant in character and which held no particular technical or organizational expertise to counter the actions Stokes was able to take.

The early history of the Board suggests it was dominated, not only by Stokes, but also by one of its members, who was a strong supporter of Stokes. This Board member was Adrien Falk, an aging San Francisco businessman, who was keenly interested in making BART a reality. Falk became BART's first Board President and in that capacity worked closely with Stokes in Sacramento to reduce the percentage majority for bond passage. Falk was a strong leader who earned the respect of his fellow board members and whose most important role was to be the major bridge between the Board and Stokes and the BART organization.

After San Mateo and Marin counties dropped from the District, Board membership dropped from 16 to 12. This cutback, coupled with normal turnover on the Board, left Falk, despite his age, a central and powerful figure. Since

Falk was a Stokes man, who approved the style and direction of Stokes' leadership, this strong pair of men tended to dominate and control the Board in BART's early years.

Asked how he would characterize the Board in terms of its technical and other skills, long-time Board member Silliman says:

> It was a mixture of the usual attorneys and lay people. But there was something wrong with our Board. I don't know whether I can say this correctly. We would meet, and maybe this is not good government, but we would disperse and as a group we had no connection with each other. We'd come here and we had all this material fired out to us. I think we did a really good job digesting it, holding the meetings, and doing these things. The laws are against it (I know the Brown Act here in California prevents privacy), but there comes a time when you want to sit down and make a family decision.

Here Silliman simply questions the Board's ability to work out its problems outside of formal board meetings, at which decisions emerged from formal, open discussion and formal votes. Politically sophisticated observers know that action taken in this way is not necessarily the same as that which follows more closed, informal and candid give and take among the members of a governing body. As a matter of fact, in retrospect it appears as though the leadership of Adrien Falk and some of the directions in which it took the Board might well have been unacceptable to the Board had it been able to caucus and know its members' true will. And Stokes, as a replacement for John Pierce, BART's first general manager, might perhaps have been less acceptable than would a technical man, simply because the Board itself lacked the technical expertise needed for some of its policy and decision making.

"But we left things up to Adrien and Bill. They sort of ran the District," says Silliman, who, in retrospect, clearly feels the Board should not have left things to Adrien and Bill.

The general tone of the early Board was that they were acting as strangers, unfamiliar with each other, unused to their role and diffident in their use of the power which the Board role conferred. Naturally enough, that power gravitated to those ready to use it, and that meant Adrien Falk and Bill Stokes.

Financial and Technical Difficulties

But it was more than personalities which influenced the shape of the developing BART organization. A number of events also had profound effects. Within a week of the successful passage of the bond issue, a taxpayer suit was filed against BART in Contra Costa County, challenging the legality of the BART District and its contract with the three joint venture firms. Although the suit was settled in BART's favor, it had the effect of delaying the initiation of work by a full six months. Six months was later to stretch to nearly three years from planned to actual start of revenue service. And during those years of delay, inflation relentlessly gouged away the purchasing power of the original $792 million and its nearly $.50 rate on property taxes that is a legacy to taxpayers until the next century.

As a technical enterprise, BART's history can be divided into three major parts: the construction of the stations and rail systems, the development of the cars themselves, and the design and installation of the control system electronics. The latter two parts are intimately connected, and in their development and testing lies the issue with which we are concerned. It was the construction phase, however, which brought to the BART organization many of its chief engineers and managers.

Acquisition of rights of way, laying of the rails, both above and below ground, construction, and above all, laying the Transbay Tube were truly major engineering tasks and were the first to receive wide public attention. Not surprisingly, the first technical personnel brought aboard at BART were construction engineers. Several, like David Hammond, Assistant General Manager, had a military background in the U.S. Army Corps of Engineers. The Board viewed these people as good concrete men, and the basic construction was perhaps the most successful of the three technical tasks, although it too suffered from both time and cost overruns. It is the judgement of some Board members that, as a result of the backgrounds and style of these Corps engineers, the BART organization began to take on some characteristics of a military organization. The fact that General Manager Stokes himself had no prior administrative experience, and that his only previous work besides newspaper reporting was service in the U.S. Navy might have intensified the military tone of BART. It is an easy mode for the inexperienced administrator to slip into.

Later, when Stokes claimed he had an open-door policy at BART, a director who had observed these military characteristics was to say, "I don't think the Lieutenants would go see the Major without talking to the Captain first." Implicit in this director's statement are two questions: To whom was the open door open? How well did the various echelons of the BART organization communicate with one another? From the point of view of some of BART's directors, apparently not well enough, as we shall see.

Late in the Sixties when it became evident there would be insufficient funds to purchase rolling stock, BART management and the Board had once again to go in search of funds, this time to the State and Federal governments. No more bond issues were possible, and the Bay Bridge already carried a $.50 toll owed to BART for the Transbay Tube construction. But it was obvious that without further support, BART trains would never roll on the new tracks which had been laid.

The best available answer to the funds shortages caused by delays and inflation lay in a half-cent sales tax in the three BART counties. This, however, required legislation in Sacramento, and although this taxing authority was finally obtained in 1969, it came at a high price. For it, the local governments gave up their "full rights to ownership," and with it came one more political body --the State Legislature-- with the "right" to review BART.

By the late Sixties all of the major tasks--construction, cars, and a control system--were underway. The originally hoped-for 1967 date for inauguration of revenue service had come and gone, delayed by taxpayer suits, by the City of Berkeley's successful bid to put BART underground, and by the engineering problems that the choice of space-age technology choice had created.

Some claimed that the delays were also caused by the policy decision not to build a BART staff to do the design and testing work. As some Board members later perceived, it was possible that contractors, including the design consultants, made money from the delays. Or, if they did not actually make money, at least the delays were not costly to them. The price was paid by the public, because BART, as an organization, started from scratch. It had no prior experience in budget management at this level, and never developed the control mechanisms nor the contracting procedures necessary to spend with efficiency the huge sums involved. Nor, in general, was BART able to develop the administrative and management procedures which would match the complexity of its technology.

Out of these events naturally grew dissatisfaction. The glamour was gone and large segments of the public felt it was being taken for a ride--and not on the BART system. These public attitudes were increasingly reflected among the directors, some of whom were now more openly questioning Stokes. Management's broken promises and its incessant demands for engineering changes and more money alienated some of the directors who had formerly stood foursquare behind Stokes.

As the decade of the Seventies opened, still without an operating system, the Board found its position increasingly uncomfortable. It was still dominated by staunch Stokes supporters, but more and more questions were being asked, and it was very clear that a number of directors were dissatisfied with BART management and, indeed, with the control the Board itself exercised over the system.

It is worth noting that the Board operated primarily through a committee structure, with the full Board approving or disapproving committee or management recommendations. BART personnel served as staff to these committees, and so it became very important to Stokes to serve these committees well. In the view of some opponents of Stokes, he also served them selectively. If you were not a Stokes supporter, you heard very little indeed except at official meetings.

Another aspect of the enabling legislation which discouraged the Board from the necessary intensive involvement in BART's affairs involved Board members' remuneration. In addition to expenses, each director received $50 per meeting up to a maximum of five meetings per month, including committee meetings. Beyond five meetings, there was no additional pay. As a result, it is reported, attendance tended to drop sharply after the fifth meeting. This factor, coupled with their authority to prepare and, if necessary, to change the agenda, left Stokes and BART management with a great deal of power in making policy decisions. It was not difficult to schedule sensitive and controversial issues for those meetings likely to be attended by fewer members of the Board, and hence to minimize opposition to management's recommendations.

From the late Fifties, which witnessed the enthusiasm surrounding the birth of BART, to the late Sixties, with its growing disillusionment, we see a Board slowly maturing to its responsibilities. For most of this period, the Board was divided, unsure that it really knew what was going to happen. Some members questioned the wisdom of the decisions that placed so much power and control in the hands of BART's consultants and contractors. Others questioned the power left to the General Manager, and as important, began to express reservations concerning the Board's interpretation of its own powers to control management. Throughout the Sixties the Board followed its General Manager's lead and was reasonably docile in the exercise of its authority. But this pattern was beginning to change as BART entered the Seventies. Not all was well at the Board level, and it would get worse in late 1971, when the scene would be dramatically changed by the actions of a group of BART employees, their supporters outside the organization and a new and aggressive member of the Board of Directors.

Chapter 3

The Grandness of the Dream:

Management's View

When people build new organizations they must decide upon the structures they will use to allocate tasks, ensure communication, coordinate activities, and distribute decision-making authority. Some organizational analysts believe that the top management of an organization has a great many options in selecting a structure best suited to the goals or purposes of the organization. Other analysts, however, believe that the particular technology, goal, or employee composition (e.g., density of professionals) determines the kind of organization structure that must be established. Thus, they would argue that organizations using routine technologies and well established production systems, and employing few scientific or technical professionals can appropriately use a classical pyramidal structure and centralized authority system, resulting in a "top-down" flow of information and decision-making. On the other hand, organizations using complex and less well codified technologies, and employing many professionals, are constrained to build "flatter" organizations with more decentralized authority and greater participation in decision-making by professional employees.

The choices of internal structure made during an organization's formative period can profoundly affect the degree to which the organization attains its institutional goals and maintains harmony and cooperation among its members.

Daniel Burnham, the architect of Chicago's magnificent lakeshore, once said, "Make no little plans; they have no magic to stir men's blood." Whatever else BART--the Bay Area Rapid Transit System--has been or done, it has had this special magic for those responsible for its design and development.

Those who managed this transportation enterprise from the time when it was no more than a dream until now--when it is an operating reality--all felt a little touched by this magic. They shared in a vision which had been glimpsed as far back as the early years of this century. Beginning back in 1911, the idea of a tube under the waters of San Francisco Bay had captured the imagination of forward-looking transportation planners. In fact, this concept--a tunnel designed exclusively for the movement of high-speed electric trains between Oakland on the east and San Francisco on the west--was almost inevitably developed by anyone familiar with the physical geography of the Bay Area, with the evolution of its commercial and residential areas and, in more recent years, with the limitations of the Bay bridges and their highway approaches in handling the welling tide of automotive traffic flowing into and out of San Francisco.

The city, a blunt-ended thumb surrounded on three sides by water, is a center of tourism and of commerce for the nine-county area surrounding the Bay. Yet, almost by definition, a peninsula poses intense transportation problems for those wishing to enter or leave it. To help solve these problems, the original architects of BART dreamed of a system which would swiftly, silently and safely move thousands of passengers a day through a trans-bay tube in bright, clean and comfortable electric trains at a rate 2 1/2 times greater than would be possible in automobiles on a cross-bay bridge.

In addition to whisking its passengers under the water separating San Francisco and Oakland, such a system could reach out to serve all the spreading cities and suburban communities of the Bay Area. Those who originally foresaw the need for a system like BART could hardly have anticipated either the magnitude of growth of population in the Bay Area during the last 20 years or the effects of the development of the rail system in touching off a wave of business construction in downtown San Francisco. The rising value of property in outlying areas as well as in the heart of the city is, in no small degree, a consequence of their proximity to BART service.

A spiritual ancestor of Daniel Burnham was Major Pierre Charles L'Enfant, the 18th Century designer of the master plan for the District of Columbia. "We must not start with a small and mean plan," said L'Enfant, "for out of it will come a small and mean city." The planners of BART shared L'Enfant's philosophy. Although the final shape of the BART system is not as grand and its impact not as great or as widespread as was originally contemplated and hoped for, it is still a major achievement in the history of American urban mass transportation.

The 1947 recommendation of a joint Army-Navy Review Board to construct a Transbay Tube for the exclusive use of a high-speed train system was supported by the final report of the San Francisco Bay Area Rapid Transit Commission. In 1953, after six years of the most intense study and deliberation, this 26-member Commission, created by the California State Legislature, advised the formation of a five-county rapid transit district which would coordinate its plans with the Bay Area's total plan for economic and social development. The prime responsibility of the rapid transit district would be to build and operate a high-speed rail system to link together the major residential and commercial centers of the area. The recommendation was turned into reality with the creation, in 1957, of the Bay Area Rapid Transit District by the California State Legislature. The embryonic BART management team was assigned the task of guiding this endeavor through the political, economic and technological thickets which stood between the dream and the reality.

Unfortunately, metropolitan areas in America today are jungles of unplanned and overlapping levels of government: state, city, county, township--to say nothing of assorted regional commissions, regulatory bureaus, port authorities and a hundred and one others. To establish any comprehensive public transportation system in the face of such a maze of governments--each jealously guarding is petty authority and self-interest and each equipped with its quota of bickering bureaucrats--reminds one immediately of Dr. Samuel Johnson's comment about a dog walking on its hind legs: "It is not done well but you are surprised to find it done at all." That BART was able, no only to do it at all, but to do it so well that, in many respects, it is a model of beauty, effective environmental planning, speed and safety, is ample testimony to both the creative imagination and the down-to-earth pragmatism of the designers, administrators and engineers who supervised planning and construction of the system through more than a decade of political, economic and technological hardship.

The Launching of the Projects

At first the task seemed monumental. The State Legislature had to be lobbied, first, to create the five-county San Francisco Bay Area Rapid Transit District, and then to grant the District the power to seek approval of general obligation bond issues and to levy property taxes to pay off the principal and interest of the bonds and to cover other specified costs.

Simultaneously, a small and talented engineering staff had to be drawn together which could develop preliminary engineering concepts for the system, and hold literally hundreds upon hundreds of meetings in communities throughout the five-county district to urge citizen contribution to preliminary consideration of the various routes and station locations. For five years, from 1957 to 1962, the small management group did all of this essential but unrewarding groundwork which was required before the first full-scale engineering design for the system could be begun.

The frustrations experienced by this group are almost inconceivable. Political and economic objections were raised by local groups and political leaders for a variety of reasons, both of principle and self-service. And then, after some five years of providing information, answering questions and objections and cultivating local sentiment, BART was finally unable to hold together the original five-county District. At the end of 1961 the supervisors of San Mateo County--south of San Francisco--withdrew their county from the District on the grounds that it would increase the county property tax too much and that the Southern Pacific Commuter System was serving the county adequately. Within five months, the District suffered another defection: Marin County, across the Golden Gate Bridge to the north of San Francisco, also withdrew from the system because of an inadequate tax base and because of an unresolved engineering question of whether the Golden Gate Bridge could handle BART trains coming from the north.

In spite of these losses, much remained. The city of San Francisco and the densely populated East Bay counties of Alameda and Contra Costa, the heart of the plan, were still in the District. In 1962, slightly more than the necessary 60% of the voters of the District approved a three-county BART plan and the sale of a $792 million bond issue to support it. With additional revenues to come from the California Toll Bridge Authority and from bonds issued against future operating revenues, the total projected cost of the system was $996 million. It would be the largest single public works project ever undertaken in this country at the local governmental level. With the elimination in 1962 of a final obstacle--a taxpayer's suit against the District which cost six months of time and $12 million in delay--BART was on its way. The importance of the BART system as a trail-blazing effort in the field of public transportation was underscored by the presence in June, 1964 of President Lyndon B. Johnson at the start of BART construction--a groundbreaking ceremony for the Diablo Test Track in Contra Costa County.

How did BART organize itself for this monumental engineering task? As in every other aspect of this effort, management tried in the early sixties to take the long view--to look forward to the time when the planning and construction phases would be over, and when revenue service would be initiated. The level and even the type of personnel and of effort required for these different phases of the project were very different. In the words of former General Manager B. R. Stokes,

> We did not want to build a huge staff for the construction effort, that would later have to be disbanded...The decision was made to rely very heavily on...consultants in the various disciplines and to only build a staff based on largely permanent needs as we transitioned into an operating status.

BART recruited, therefore, a small and highly talented staff of engineers and supervisors who would be so intimately involved in the system from the very beginning that they could carry over effectively and essentially intact into the operating phase. If it had done otherwise--if it had hired directly the additional hundreds of engineers and technicians needed for the design and construction phases of the project, the dislocation would have been enormous at the time of transition from construction to operation. As a result of this recruiting decision, therefore, the BART staff itself was fairly small and compact and was able to interact across group and individual lines with considerable freedom and informality. In addition, the management staff perceived itself as a permanent group with a long-term commitment to both the construction and the operation of the BART system. The validity of this perception is underscored by the low rate of turnover experienced at the BART managerial level. Erland Tillman, Director of Engineering and Construction, has, as a matter of fact, expressed his pride at the fact that the group which he helped recruit was extremely stable from the time of their employment forward.

There was no one method of recruitment of managerial or technical staff. Tillman notes that:

> Engineers for the BART staff were recruited by a number of means. These include formal advertising, personal contact with other engineers who were in a position to know of people who were suitably qualified and who were available, and personal contact with persons known to be available.

B. R. Stokes adds that all possible recruiting means were used, since the system's goal was simply to attract the best and most competent staff it could.

From the beginning, the expressed goal of all the managers of BART, in the face of a plethora of political, economic and technical problems, was to build slowly and deliberately a stable and harmonious staff of the most highly qualified engineers available. These individuals were brought in with a variety of competencies: specialized knowledge of certain key areas, broad and general understanding of the wide range of engineering problems which would be encountered, and general management abilities, especially when these had been demonstrated in direction of complex and large scale systems. It was only with this personnel policy, top management reasoned, that the whole grand conception could be seen through to a successful conclusion.

An important factor in developing and maintaining organizational harmony was the open-door policy enunciated by and apparently adhered to by virtually all of the top managers of BART. Erland Tillman, for example, has said,

> I operate on the basis of dealing with various individuals at all levels, depending upon the problem and more or less on an open-door policy. I believe that you hold responsibility through the chain of command, but that you talk to everybody in the chain of command and get to know them and operate on the basis of getting them to know you and feeling free to come and talk to you.

In the same way, David Hammond, Assistant General Manager for Engineering and Operations, has characterized his own managerial style in these words:

> My style of management is to go to the guy who's got the problem or got the information. I don't mean that I would be bypassing Tillman and Ray (his two principal subordinates: Director of Engineering and Construction and Director of Operations), but (when there was) the sort of thing where I needed to go directly to somebody, I did; and in turn, there were a fair number of people on their staffs who had direct access to me.

General Manager B. R. Stokes says that over a period of time, many employees come directly to see him,

> from telephone operators to line operations people to members of consulting firms...Some people operate on a basis of kind of remaining in an ivory tower. I don't. I try to get out and know as many of the people employed as I can. And so I was out on the job and out on construction a lot and people at all stations in the organization came in on occasion.

The Creation of the PBTB Consortium

But how could management simultaneously solve the conflicting problems of keeping BART's own professional staff relatively small and permanent, and yet of having the best available engineering talent working on the design and construction of the system? It decided on the technique of contracting with a consortium of outstanding engineering firms and having them serve as its basic designer of system facilities and equipment, its supervisor of construction and its monitor of the major contractors of the system. The three firms--Parsons, Brinckerhoff, Hall and MacDonald, of New York, Tudor Engineering of San Francisco and the Bechtel Corporation of San Francisco joined together to form the organization called PBTB--Parsons, Brinckerhoff, Tudor and Bechtel. This consortium did the basic design of the system. A successor to Parsons, Brinckerhoff, Hall and MacDonald-- Parsons, Brinckerhoff, Quade and Douglas, later combined with Tudor and Bechtel as general contractor during the construction phase of the system. It is unlikely that a more impressive array of engineering talent has ever been assembled anywhere for the direction of a major public works project.

The criticism has been leveled that the PBTB consortium--as well as BART's own engineering staff--was long on construction talent, but short on engineers capable of monitoring the complex electronic and computer technology of the train control system. The claim is made that the effective completion of the concrete parts of the BART system as compared to the problems in train control confirms this judgment. But management insists that this is not so. Erland Tillman, for example, says that, in addition to its engineers with strong backgrounds in civil and construction engineering, BART had others "with strong backgrounds in electric and electronic disciplines as well as those with strong backgrounds in transit and railroad rolling stock equipment." It may simply be that the concrete parts of the system were grounded much more in conventional technologies than were the complex and innovative electronic and computer technology of the train control system.

The PBTB consortium shared this breadth of professional expertise. B. R. Stokes notes that those who suggest that PBTB had real engineering strength only in the construction phase of the project must be unaware of the Bechtel Corporation's long and extensive history of development and installation of highly sophisticated electronic control systems in nuclear, hydro-electric and other complex modern installations.

Finally, the charge that the presence of former Army Corps of Engineers officers at various executive levels in BART lent the organization a tone of military authoritarianism is dismissed as sheer nonsense by the former General Manager. Indeed, B. R. Stokes has expressed thankfulness that BART was able to recruit so many former Corps engineers since they, and their Navy counterparts, were "the best and most experienced people available for the management of complex, large-scale systems, and BART was the biggest public works system in history." (telephoned comment 12/77).

The in-house expertise of BART's own staff, in combination with the competence of the PBTB consortium, was reinforced, according to Stokes, by a Board of Consultants, drawn from among top public transit authorities in cities all over the world. This board met periodically to give advice on staffing and on program development. In addition, BART had on call a large technical advisory group which met two to three times a year to review all the operations of the developing system: staffing, construction, control system, test track operations. The total body of high-level technical opinion available to BART management is probably unparalleled in the history of large-scale public works design and development.

With the decision to contract with the PBTB consortium as consulting and supervising engineer for the system, the intended policy of BART's management that professionals who came to work for the system stay on for extended periods--well into the revenue phase--was fulfilled in practice. The results were just what had been hoped for. All of the system's top-level managers testify that, by industry standards, BART experienced a very low rate of personnel turnover and that employment was at all times very stable up and down the professional ranks. According to the top administrators of the system, the relatively small size of the staff and the open administrative style of most of the managers undoubtedly had a great deal to do with this. But there was another reason: there was a tone of excitement and adventure at BART. Everyone felt himself part of a really trail-blazing effort, and there was, in the words of B. R. Stokes, "a kind of glamour about the project."

BART's Organizational Style

Because they felt part of a team, most BART people were apparently successful at working within the organization to get what they wanted--or what they perceived was needed for the system. This was what was so unusual about the case we are dealing with: it appears from management's point of view to be a question of three employees not judging adequately or grasping the possibilities of working successfully through organization channels. Erland Tillman says that, in general, there were no organizational roadblocks in the way of those employees who wanted to raise questions through the designated channels of BART.

As a matter of fact, it was really part of the obligation and responsibility of all BART's professionals at all levels to raise questions--to look for problems. Since they did not have the specific responsibilities of directing construction of the system-- this being the job of PBTB--the usefulness of BART engineers was almost directly proportional to the questions and issues which they raised. Assignments of BART engineers were not, in general, cut and dried and limited in scope. Rather the District wanted its engineers to have free rein to explore and to make inquiries on their own. David Hammond, Assistant General Manager has emphasized this point in saying that

> the whole system was for the staff to be analyzing what the problems were--more than just tabulating problems--and (for them to be) coming up with what actions were necessary to resolve the problems.

BART, then, according to its top management, ran an open system in which professionals were encouraged to behave professionally--to exercise creativity, ingenuity and independence of judgment within the organizational setting in order to help management and the Board understand better what the real issues and problems were in this admittedly novel transportation system.

Charles Kramer, Superintendent of Power and Way, reinforces this idea in specific terms as he discusses the interaction of members of the track and structures group, the buildings and grounds group and the electrification group, all of which reported to him. Kramer was asked whether any procedures existed for having these groups interact with one another without going through him, since it was assumed that there would be a great deal of wasted time and effort if all messages had to go up and down in order to move laterally. Kramer replied: "There were no formal procedures...They were all expected to interface and work together."

People were expected--if they needed to communicate with someone in another group or section--simply to go directly to that person and try to solve the problem at hand. In the same way, supervisors tended to spend a great deal of time simply talking to and exchanging information informally with BART workers at all levels--simply to maintain a good working contact.

"This informal interchange between people within the department was not only expected, but required, because they can't function as individuals..." says Kramer.

He goes on to respond very positively when asked: "There was no requirement, then, that they (engineers) would have to go up the lines of command and down the lines of command?" by saying: "None whatsoever!"

At another time John Fendel, in charge of construction engineering, states:

> I didn't have any chain of command really within (my) group. I didn't use my assistant as a chain of command. I had weekly meetings with all my people at which time they were all to bring up the problems they were working on and any ones they considered they needed help on. The objective, of course, of the meetings was so that we didn't invent the wheel three times; so we would have a problem and solve it one place and make sure that everybody understand that that was the way we solved it here.

Indeed, one of the three engineers with whom this study is concerned, Robert Bruder himself, has some things to say about the general organizational climate at BART. Bruder was asked about Fendel's operating styles with his group. He replied:

> The whole group met on Friday for a couple of hours from 10:00 to 12:00...and we would take turns explaining to each other how our contract and how our phase of work were going...just as general interest and interface and Mr. Fendel would take notes and then on Monday he would go to his staff meetings...It was his way instead of a bunch of written reports every week...
>
> We were expected basically to be out in the field at least over half the time...so that we were aware of what was going on and meet the other PBTB people that were the project engineers either at a station or in the systemwide office.

Questioned as to whether the weekly meetings were effective, Bruder replied: "Yea, it filled all of us in." Asked whether his supervisor (Fendel) was generally accessible--apart from the weekly meetings--Bruder said: "Oh yea, he was always accessible." And of his associates with whom he worked in other areas he said: "They were real pros".

Finally, Bruder was asked whether BART had a favorable, positive organizational climate and he replied: "Yea, yea...everybody knew his field and he felt confident, and there was freedom within the group to go back and forth and to ask questions if something came up and you needed support."

Perhaps the point of openness and informality is belabored a little too much, especially in view of the many assertions which were later to be made to the contrary. But structural openness is one of BART management's chief points of pride. The chief administrators of the system claim that, precisely because BART was operating on the cutting edge of technology in developing a modern mass transit system, they simply could not afford _not_ to be open, because ideas and solutions could come--and literally did come--from all levels of the organization. Of course, the BART staff was enormously dependent on its contractors--Rohr for the cars, Westinghouse for the Automatic Train Control system, IBM for the computerized fare collection system--and on PBTB for overall contracting and engineering supervision. But the BART professionals appear to have interacted continually with the designers and engineers of these organizations, and many of the specifics of train design, automatic train control and all the rest were the product of cross-fertilization of minds in all the organizations. Management so much wanted its own people to be in the middle of things so that it even sent one of the engineers involved in this case, Holger Hjortsvang, to Pittsburgh for the better part of a year to observe and work with the Westinghouse engineers who were designing the Automatic Train Control System. The men who would later be on the operating firing line had to understand the complex workings of the new technology BART was contracting for and using.

BART management had, really, four goals for this system: safety, speed, reliability and comfort, and it could not afford to slight any of them in favor of the others. But if there was one highest priority, it had to be, as is true of any mass transit system, safety. As a matter of fact, when the system began revenue operation, it did suffer to some degree from a lack of reliability--and that was occasioned only by management's concern for safety. It was so necessary to be alert to any possible hazards that the overall service was less reliable than it might have been.

During the design and construction phases of the project, BART's management truly believed--and still believes--that it ran an administratively effective and technologically forward-looking organization which was open to new ideas and to constructive criticism, whatever the source. On the other hand, the evidence makes it clear that BART was not some kind of technological and administrative Garden of Eden. It had its internal problems, as well as the enormous difficulties imposed on the organization from without.

These problems were technical and organizational as well as political and financial. Nor was the system perfect in the matter of personnel relations. No organization is. It had some unhappy employees. It had its share of petty grievances. But in comparing the general level of morale throughout most of the history of BART with that which exists in other organizations--industrial or governmental--BART appears to come off pretty well. Most of the people were enthusiastic about the goals of BART while they were being recruited--and most of them appear to have liked what they found after they joined. BART managers tended not to be sticklers for protocol--and they gave their professionals the degree of freedom and independence which professionals deserve. There were, to be sure, ungodly pressures on the Board and on management: taxpayers suits, high bids, inflationary strikes by suppliers employees, legislative delays in getting additional funding and, apparently funny but costly, gopher attacks on the insulation of control cables. There were goals not achieved and deadlines missed. And the impact of these problems and traumas undoubtedly affected managerial temper and temperament. But through most of its construction phase BART was a healthy organization whose members knew what they were doing, and who were doing pretty much what they wanted to.

Construction Begins

In January, 1966, eighteen months after President Lyndon B. Johnson officiated at the groundbreaking for the Diablo Test Track in Concord, the basic construction of the BART railbed began. The system, combining elevated, surface and sub-surface components, faced its greatest engineering challenge in the construction of the Transbay tube. Work on the tube began in November of 1966 and was completed just three years later. At the same time that this remarkable feat of civil engineering was being carried forward, other construction crews were building elevated tracks on the Berkeley--Richmond, Central Contra Costa and Southern Alameda lines, were designing and digging the tunnel through the East Bay hills, and were tunneling through mud and a one-hundred-year accumulation of underground utility lines and ducts to construct the Market Street subway and stations in San Francisco itself. The pressure of the mud and San Francisco's high water table made it necessary to do the six-year-long subway construction, 80 to 100 feet below the city's bustle, entirely under compressed air conditions. PBTB's engineers had to use special monitoring equipment to protect the safety of the BART sandhogs and BART developed its own medical center to meet their health and safety needs.

While none of this, of course is strictly relevant to the specific issue of engineering ethics which will later be addressed, in a general sense, it does bear on the question of the engineering ability and the professional integrity of the BART staff and its PBTB consultants. It is noteworthy that the Transbay tube and the entire Market Street subway complex were, at the same time, award-winning civil engineering accomplishments and models of safety in heavy construction. To have built these structures under the incredibly difficult and dangerous conditions which were faced and to have done it with one of the best safety records ever achieved in comparable civil engineering projects suggests two things: one--BART had working for it one of the most competent assemblages of engineers ever drawn together for a major public works project, and two--no project was ever undertaken by BART without giving the highest priority to the safety and the well-being of the people involved, whether they were the system's builders or users.

At the same time that the tunnels and elevated structures were being constructed, BART was contracting for the system's rolling stock and control system. In 1967 a contract was awarded to the Westinghouse Electric Corporation to design and build the Automatic Train Control System (ATC), and in 1969, Rohr Industries received the contract to supply the first 250 railroad cars to be used on the BART system.

Tribulations of BART's Managers

The achievement of basic construction of the line and the award of the two prime equipment contracts were accomplished against a background of intense political and economic turmoil and tribulation. Management seemed always to find itself in an intolerable squeeze between inadequate funding on the one side, and, on the other, the continuing demands of the public and their political representatives for modifications of and additions to the original design of the system. In other words, the continuing theme of BART development was the public demand for more and more, but with the insistent provision that it cost less and less.

Speaking realistically, however, it is difficult to understand how a project budgeted at almost one billion dollars can be said to be inadequately funded. The problem, from management's point of view, was not in the initial level of funding. Rather, it is what the repeated delays and unanticipated rate of inflation did to the original estimates. Start with inflation. The original estimates projected an annual inflation rate of 3-1/2 to 4%. During the principal years of construction, BART was faced with an average annual inflation rate of 7%. On a base the size of BART's, a 7% inflation rate is almost ruinous. At the same time, management was beleagured by a host of private interests, each insisting that the environmental impact of BART on their community or neighborhood be ameliorated, or that the system be redesigned to meet their special needs. For example, well after rights-of-way had been obtained and station locations provided for, local pressure groups forced relocation of 15 separate BART stations and a total of 15 miles of right-of-way.

In Berkeley, the city reversed an earlier approval of a combination aerial and subway line, and pushed for a subway--only line through the city, a very much more expensive option. This meant a total redesign of the Ashby station, precipitating a controversy which lasted 2-1/2 years. After the resolution of this issue, the cost of which to the total project in terms of delay as well as basically more expensive facilities is almost incalculable, a Berkeley City Councilman filed a successful suit to force a second redesign of the Ashby station.

The management of a complex and multifaceted enterprise like BART, which demands coordination of the contributions of engineers and technicians with a host of different specialties and skills, and smooth integration of the varied products of their effort, is extraordinarily difficult, even assuming the absence of economic pressure and social and political demands. To manage such an enterprise in the face of continual badgering and harassment--to say nothing of downright opposition and obstructionism--takes the skill of a blindfolded tightrope walker juggling six balls as he inches across the chasm.

Yet, BART managed to survive each crisis. In 1968 the District's Board applied for and received a $28 million grant from the Department of Transportation to help purchase rolling stock. The following year, the California State Legislature approved a BART District-wide sales tax of one-half cent to provide another $150 million needed to complete the system. With this approval, BART's managers knew they would achieve their goal.

Chapter 4

Working for BART:

The Engineers' Experiences

People join organizations for diverse reasons. Those who are forced to join organizations by virtue of being imprisoned, drafted or legally committed are likely to be motivated to act on behalf of the organization only by threat of coercion or deprivation. Those who join organizations largely because of attractive remuneration possibilities, will be motivated primarily by opportunities to increase financial rewards. But those who are attracted to an organization because they identify with it and its goals may not need the promise or threat of external reward or punishment to get them to work in its behalf. These will be eager and willing workers. In fact, it is often difficult to direct and restrain the activities of these strongly committed organization members.

Among those who bring to their work strong professional, idealogical or moral commitments may be professionals employed in large organizations. When such commitments are consonant with organizational action, the result will likely be a satisfied and contributing organization member. When they are not consonant, however, a basis develops for independent views and assessments by these professional employees of an organization's policies and actions, and for possible intra-organizational conflict.

In addition, the degree to which an organization gives its professional employees appropriate professional responsibility and authority can also serve to limit or enhance these em-

> ployees' opportunities to achieve personal and professional goals. A professional who identifies with his organization and its goals may resist accepting a definition of his position which emphasizes narrow specialization and fragmentation of tasks, and may instead seek to participate in discussions and decision-making related to the broad goals of the organization.

Thomas Hardy's poem, "The Convergence of the Twain" tells something of the history of H. M. S. Titanic and of the fatal iceberg which caused her tragic plunge to the floor of the Atlantic. During the very time that the great ship was being built, Hardy writes, in shadowy, silent reaches the iceberg also grew. But the two seemed so alien and so distant in birth that

> No mortal eye could see
> The intimate welding of their history,
>
> Or sign that they were bent
> By paths coincident
> On being anon twin halves of one august event.

In much the same way, during the decade and more that the BART system was being planned and built, three engineers were developing their three unique personal and professional careers with no real sign that they and the system would likewise be "twin halves of one august event." But was there more in their histories than in those of the ship and the iceberg to predispose them to the ultimate collision? A look at their histories, even if it does not fully answer this question, may provide a better understanding of their behavior.

Social and Educational Background

Holger Hjortsvang was born in 1911 in Copenhagen, Denmark. He came to the United States in 1956 from Venezuela, where he had been employed by the Mene Grande Oil Company, a subsidiary of the Gulf Oil Company. He decided to stay in this country, and became a naturalized U.S. citizen in 1963. He is married and has three grown children, two girls and a boy. The girls are married and, at last report, his son was serving in the Coast Guard.

Hjortsvang earned a diploma in electrical engineering from the Institute of Technology in Copenhagen in 1936. While he has no advanced degree, he took evening extension courses in electrical engineering at the University of California at Berkeley during the four years from 1956 to 1960. He again enrolled for formal courses in the early 1970's in preparation for the State Examination of Professional Engineers, which he took in June, 1973.

Robert Bruder was born in 1923 in Aberdeen, South Dakota. He has a twin sister and two younger brothers. His father was a traveling salesman dealing with wholesale groceries.

When Bruder was about one year old, he and his family moved to northern Michigan, where he grew up and graduated from a parochial high school in 1940. Bruder describes himself as a good student in high school, with real strengths in science and math. He won a scholarship for a major part of his college expenses and entered the Michigan School of Mines, where he majored in electrical engineering.

After finishing two years at the School of Mines, Bruder joined the Navy. He went to boot camp in Idaho and was then sent to Texas A & M near Bryan, Texas, to get training in electronics. It appears that Bruder had some problems in the Navy, stemming from his wish to transfer out of electronics school and to enter the Chaplain Corps. His attempts to transfer created substantial difficulties with the Navy, and he was discharged from the service before he had completed a year of duty.

Bruder's father died while he was in the service, and his mother moved from Michigan to California. After being discharged from the Navy, Bruder joined his mother and worked at a number of routine jobs. Among other possibilities he considered getting a radio license and joining the Merchant Marine, but finally decided to enter a Jesuit seminary and study for the priesthood. In late 1943, therefore, he entered Jesuit Seminary in Los Gatos, California.

After 3 1/2 years in the seminary, Bruder decided that the priesthood was, after all, not what he really wanted, and so he returned to his technical interests, without, however, fully resolving his competing interests in engineering on the one side and philosophy-theology on the other. In 1949, he entered Loyola University in Los Angeles to complete his studies in engineering.

Bruder was an active student who was a member of a ski club and the German club, and was a participant in intramural sports. In contrast to his strong academic record in earlier college days in Michigan, Bruder describes his performance at Loyola as average ("I was not a prize student"). He attributes this to his age and realism about why he was in college ("...getting a piece of paper...that's what the world wanted.")

Bruder received his bachelor's degree from Loyola University, Los Angeles, in 1953 and began his career as an engineer. Today he is married, the father of four children, and still a practicing engineer. In reflecting on his life since high school days, Bruder affirms his continuing interest in religion and philosophy: "My profession is just a way to make a living. From my total background in education and everything else I'm...more a philosopher than an engineer."

Max Blankenzee was born in 1941 in Surabaja, Indonesia. He lived there for eight years with his mother, an older brother and stepfather, who was a military man and small businessman. In 1950, the Blankenzee family moved to Holland, and the family grew with the addition of two more children, a boy and a girl. As a youth, Blankenzee was interested in designing and inventing things, and was a very successful student in a technical high school. Because of his interests and scholastic achievements, he entered the Aeronautical Institute in the Hague.

In 1961, one year after receiving his degree in engineering from the Institute, Blankenzee emigrated to the United States. Because of language problems, his youth, and the fact that he was not a citizen, he was at first unable to find work in his professional area. As a result, he was forced to work as a garbage man, supply man, and a junior chef. In 1964, he became a naturalized citizen and in 1965 he obtained his first engineering job, with Varian Associates.

Early Work Experiences

Holger Hjortsvang

When Holger Hjortsvang came to the United States in 1956, he first worked as an applications engineer for Lenkurt Electric Company in San Carlos, California. He describes his work:

> (I would) prepare the specification for a customer according to his order. The customers were mostly telephone companies, and they would order carrier equipment for their exchanges, and these would consist of so-and-so many lines and so many subscribed connections, and so on. I was laid off in December of 1957 because there was a business recession at the time and hundreds of engineers were let go.

Hjortsvang was concerned about having to travel too much on this job and being required to be away from his family for as much ten months when he was assigned to an installation in Muskegon, Michigan. "Even if I had not been laid off, I would have considered finding some other work," he said. Despite his reservations about the position, Hjortsvang's employers, according to him, were satisfied with his work: "They told me that they would be willing to rehire me when times got better."

Immediately upon leaving Lenkurt Holger took a position with Lynch Carrier systems for about three months, leaving in April, 1958. He gave a week's notice to Lynch when "I got an offer for considerably better pay from Shand Jurs Company." He was with Shand Jurs from April, 1958 to June, 1962, working on the "development of a supervisory control system, solid state system...for use in the petroleum industry, used in connection with tank forms, pipelines and refineries."

When asked about why he left Shand Jurs, Hjortsvang replied:

> Well, the company had gotten in financial difficulties and had been bought up by General Precision, Incorporated, and it looked like the operations would be transferred to Chicago. And I didn't really like to move from this area, so I looked for another job... I was considered a very competent designer, and I was told so. (My supervisor) had on many occasions expressed his admiration for what I did.

Hjortsvang next took a position with Beckman and Whitley from June, 1962 to October, 1963.

> I was again a sort of systems engineer. My title was project engineer...it was the same kind of work I did at Lynch and Lenkurt, essentially assembling systems to customer specifications. Now in this case it was meteorological systems; systems that would measure and record wind velocity, temperature, that sort of data.

Why did Hjortsvang leave Beckman and Whitley? Apparently the same story: financial trouble. The company didn't sell enough of their meteorological systems, and they laid off one half of their staff in Hjortsvang's department. When questioned about problems he is alleged to have had while employed there, Hjortsvang denied that he had ever been described by his supervisor as being of average technical ability, or as being somewhat inflexible. On the contrary, Hjortsvang answered, "I know he showed quite a respect for my ability as an engineer."

After being laid off by Beckman and Whitley, Hjortsvang went to work for Nuclear Research Instruments. Hjortsvang remained with Nuclear Research Instruments until he left to work for BART on September 12, 1966.

Robert Bruder

After Robert Bruder received his bachelor's degree from Loyola in 1953, he worked for about twelve years in the aerospace industry. In order, he worked as a production sustaining engineer for General Dynamics; an engineering supervisor and senior project staff engineer for inertial guidance and digital computers at Autonetics Company of Anaheim, CA; an engineer on the Minuteman Missile inertial guidance and instrumentation launch readiness program while with the Boeing Corporation; an electronic vehicle and ground station system test engineer with Lockheed Missile and Space Corporation.

After leaving Lockheed, Bruder was employed by I.T.T. Federal Electric Corporation from October, 1965 to March, 1967. His position was Tracking Ship Manager, in which he assisted the operation manager in training and planning for Apollo Operations with a crew of 100 technicians. When he was later involved in the BART controversy, Bruder made some mention of having lost his job with Federal Electric for being a little too aggressive in trying to solve some problems. What exactly had he done? According to him, he was dismissed only for "taking the bull by the horns and trying to get rectification of problems and the work you're doing by taking steps maybe outside of normal channels." Specifically, he believes that he was terminated for going over the head of a superior.

> Well, I wasn't terminated sudden death, you know, within a month or two of taking that step; but the situation continued and I still tried to resolve what I thought was my responsibility in it. I guess I stepped out of

> bounds as far as they are concerned and they just let me go because I was too aggressive.

BART management would later try to use this information about Bruder's past employment to show that he was a troublemaker of long standing. Bruder, however, maintains that he has always followed the highest ethical standards of a professional engineer, a posture which caused him difficulty in organizations which preferred to treat him--and presumably other professionals--simply as employees.

After leaving Federal Electric, Bruder was employed with the Army Corp of Engineers at Vandenberg Air Force Base from March, 1967 to November, 1969. He was a staff specialist responsible for testing and acceptances, for power generation, transmission, substations, lighting, controls and communications in connection with the Manned Orbital Lab launch and support facilities.

Then, in November, 1969, Robert Bruder began to work for BART.

Max Blankenzee

Max Blankenzee is younger than his colleagues, and therefore, has a shorter employment history prior to the BART experience. From August, 1965 until February, 1967 he was employed with Varian Associates. He resigned that position to work Philco-Ford where he remained until he lost his job in May, 1969.

> It was a lay-off. I was told before I was laid off that they were going to have to lay me off in the next month or so, so I might as well start looking for a position. There was (nothing) on the record that says "laid off," okay? They said they would have to lay me off. It would be better if I would find another position somewhere else.

Blankenzee found a position with Hewlett-Packard as a system programmer, where he remained from May, 1969 until November, 1969.

"I left Hewlett-Packard for a better position...I wanted to get more involved in...leadership of a project, and Moore (his subsequent employer) allowed me this opportunity."

In response to questioning about how he got along with his supervisor while at Hewlett-Packard, Blankenzee has stated:

> Mr. ________ was quite involved with contractors. He could not have the time to spend with the people that he hired, and whenever it came to conversation Mr. _______ had the attitude of yelling and screaming at people in front of a whole group of people, everybody involved. I felt that that was a very low position for management. For that I never respected the man.

Did Mr. _______ ever scream at Blankenzee in front of other employees? "Yes, he did." And was this screaming in response to the quality of Blankenzee's work?

> No. He was making remarks (about) the fact that I converted some of the programs (done by) the consultants that he had hired...He had a man hired for $25 an hour who was doing a job, and at the time that the test came out, the acceptance of the thing, the system crashed. Okay? Somebody was asking me if I would know what to do to help out because it was kind of embarrassing for Hewlett-Packard in front of the Air Force. I went in there and I corrected the error and when (my supervisor) found out that I corrected the error he said that I wasn't qualified to go in there and (he) yelled at me about correcting the thing and said that...'whenever you get a doctorate behind your name, then you might be able to go in the same as this man has'.

Blankenzee was also asked if anyone at Hewlett-Packard had ever stated that he was not strong, technically, as a programmer and that he would not be successful as a programmer.

> (My supervisor) wrote those statements on my exit interview. But we parted with the agreement that those things would be taken off the list, and I was told by Mr. _____ and by the man who did the exit interview that I could come back to Hewlett-Packard and work for them again.

After Hewlett-Packard, Blankenzee was a programmer analyst for Moore Systems, a position held from November, 1969 to February, 1971, when he was laid off. "Moore went -- and lost -- everybody got cut back. There was ten percent cut in pay, and then they told me when I left that there wasn't any more work and I should go somewhere else."

Blankenzee next took a temporary position with Westinghouse as a programmer/analyst, working on some of the programs dealing with the train control system at BART. "(They) hired me on a job-shop basis to do a little bit of work testing out some of the software, and also testing out on the stations that were in the field. The position at BART came open around March (1971) and at that time I was interviewed."

Max Blankenzee joined BART as a senior programmer/analyst on May 10, 1971.

The Engineers Join BART

Holger Hjortsvang learned about employment possibilities with BART from an advertisement for engineers that appeared in the newspapers during the summer of 1966. He was interviewed by the Personnel Manager who then took him to talk with people in the organization who would be interested in his services.

> (The Personnel Manager) was mostly interested in my background. He insisted on seeing my naturalization papers and the record of my diploma to make sure I was a graduate engineer...(The Senior Maintenance Engineer) worked to determine my capability and several aspects that he thought was of interest in his group concerning train control and communica-

> tion, and he liked my varied background in communication and control systems...He mentioned that, at this time, it was a very good opportunity for advancement with BART, and he had seen an example of people coming in there, and a few months later they had been promoted to very responsible and well-paid positions.

Hjortsvang viewed his position with BART as more than just another job. He expressed a commitment to the organization's goals that he felt was special, and was largely responsible for the way in which his duties were defined in the earliest phase of his employment. As he has stated:

> I believe that the understanding between me and my superiors was that, with my very clear interest in the BART system, the technology involved and my dedication to the success of what we were doing, that I was the best person to judge what the general area of work would be for me.

Robert Bruder also learned of the possibility of a position at BART from an ad in the newspaper. He was attracted by the size of the project and by the fact that his specialty -- construction -- was a large part of it. "It was one of the most talked about projects, of course. I'd heard about it, but I hadn't really read anything in detail. Moreover, the position as described in the ad just seemed to 'fit me to a T'."

Max Blankenzee's employment with BART seemed less a decision concerning a new job, than it was a gradual transition from one organization to another without really changing his position and duties. He had been working on the BART train control system while at Westinghouse, and it was through this experience that he became introduced to and interested at BART.

While at Westinghouse, Blankenzee came into frequent contact with Holger Hjortsvang, who had been assigned by BART to work with the Westinghouse project. "I learned that there was a position open and I asked Mr. Hjortsvang if it was a possibility that I could apply for that (senior programmer). He asked me for my resume, asked me to fill out an application and he submitted it to Mr. Wargin, I think."

Next, Blankenzee had an interview with Mr. Wargin (Hjortsvang's supervisor).

> (He) told me the position was a senior programmer and that I probably (would) have to do the hardware and the software, more or less work along with Westinghouse, stay abreast of them, get involved with the development in such a way, by trying to find out what Westinghouse is doing, so that later on we could maintain the central computer complex.

The decision to work for BART reflects Blankenzee's strong interest in the goals of the organization and in its technical problems. As with Hjortsvang, working with BART was more than just another job for Blankenzee.

> What I found interesting is BART. I guess it attracted me and I thought that I could help BART out that way. But BART had no knowledge, whatsoever, in those facilities, of what happened within the central control computer and how really it was important at that time that the computer systems played a part within the operation of the train systems. Everything was the hardware out in the field, O.K.? Nobody in Westinghouse at this time (understood) the total picture the way it was. Westinghouse Electric itself was the prime contractor. One of its divisions was developing the software. Now Ken Dale was Westinghouse Electric...and he was called back to the San Francisco Bay area to monitor the software people because that other division was just doing things all by itself, all uncontrolled. And looking at what BART was up against, you know when the acceptance came up, I thought that I could provide them with that information...I'm a resident of the Bay Area. This is one of the most new innovative systems and everything that is being built about it, specially within the software, is uncontrolled. Nobody knows where they are. Nobody knows where they stand, and nobody really

> worries about it. It's a million dollar adventure...and it kind of looked like an interesting thing to tangle with.

The Job and The Organization

Holger Hjortsvang

Along with the special enthusiasm which they all had for the organization and its goals, the three engineers' early work experience at BART conveys an impression of ill defined or vaguely defined work activities and responsibilities. What appears at first glance to be an organization which provided its engineering employees with great freedom and autonomy to carry out their work, can be perceived, on closer examination as a weakly organized administrative structure which failed to provide clear and definitive work assignments. Part of the ambiguity about the duties of the engineers can be traced to the relationship among BART's engineering staff, the consulting firm PBTB (Parsons, Brinckerhoff, Tudor, and Bechtel), and the system's major contractors (especially Westinghouse). PBTB was contractually responsible (among other things) for monitoring the work of the major contractors, and the engineers were required to deal with both PBTB and Westinghouse about specific contracts involving BART'S Automatic Train Control System.

Hjortsvang describes the scope of his work activities and responsibilities when he joined BART in the following manner:

> At that time (1966-67) the organization was sort of an embryo, and the title I was assigned was Train Control and Communications Engineer..I would be assigned work by Mr. Andrews, looking into proposals, technical descriptions, be concerned with the specification of the train control system, and so on, all on a very vague basis. At the time a specification was being prepared for contractors to bid on what was called the train control and communications system for BART... A

> representative from the consulting firm (PBTB) brought their drafts of specification...and Mr. Andrews and myself went over them with this representative, commented on it.

For the next two or three months Hjortsvang would receive drafts of the specifications from PBTB, which he would read and comment on regarding clarity, additions, or corrections. He felt that there was only one aspect of these "specs" upon which he had any effect.

> That was about documentation of logic circuits...Documentation is a record of the function and the construction of these things, and also a written explanation of how it works...In the specifications, this had been omitted, and this was what I pointed out, and I wrote a letter telling what should be there, and that was accepted. It's in the specification now.

After award of the contract for the train control and communication system to Westinghouse Electric Corporation based on the specifications on which Hjortsvang had worked, Hjortsvang describes his activities:

> As soon as I could get my hands on any information from Westinghouse, I studied it in order to understand how this new system was going to work.

Hjortsvang was asked if he was directed to act in the capacity of liaison with Westinghouse after the contract was awarded.

> The interesting thing with BART is that I was actually never given any specific directives... I found that it was pretty obvious what I had to do when there was correspondence. I had to go to meetings, and so on, discuss the subjects that were up at the time with Mr. Andrews. That's what I did.

In response to questions, Hjortsvang has indicated that he was satisfied with his position at BART during the first two years.

> My time was spent studying and trying to understand the design so far as it was possible (with) the information that was available to me. I don't think at that time I was concerned about the safety of the system.

Even though he was generally satisfied with his position during these first two years with BART, Hjortsvang still perceived his role as one lacking clear specification of duties thereby permitting him considerable latitude to define his own work activities. He also perceived the structure of authority and responsibility in the organization to be informal and unclear, again giving him the opportunity to shape his own position. This ambiguity could, of course, be viewed as offering employees professional advantages as well as disadvantages. Hjortsvang's comments about his position and his relations with superiors illustrates some of these mixed feelings.

> Mr. Wargin came on board sometime after I was hired. I don't recall exactly. But he was placed as superintendent. And Mr. Andrews (Hjortsvang's supervisor at the time) was then to report to him. In other words, he sort of came in between Mr. Andrews and Mr. Aboudara. Mr. Andrews had the title of Senior Engineer, and there was no superintendent over him, and Mr. Wargin got that position.

After the employment of Wargin as Superintendent, Hjortsvang was unclear about whether or not he still reported to Mr. Andrews.

> This was never really defined. After a long time, I asked Mr. Wargin, 'Do I report to you or do I report to Mr. Andrews? Has there been any change in my relation to Mr. Andrews?' He says 'You know very well that you work for me.' He never told or published anything indicating a change in command...it typifies the information and lack of distinct definition of duties within BART, within the organization.

When Hjortsvang was questioned about his feelings of ambiguity about his duties and responsibilities and was asked why he didn't complain to supervisors about the lack of direction provided him, he stated:

> Because it was up to myself to judge what work I should do. What would be in my area, and what would not be... There's a difference between specific tasks that you don't talk with our supervisor about and say, should I do this, or he tells me, 'Will you do this task,' and the general area of work that you use your time on... The specific tasks that I talked about, they could not possibly take my whole time and energy, so there were vast amounts of time I did work as I found it would be useful for BART without any direction... No, I didn't complain, because I think that way it was in BART's best interest, and in my own interest, too.

It seems clear that Hjortsvang chose to use his time in such a way that one of the major duties of his position was "to keep track of what Westinghouse was doing." When asked if, in fact, that was one of his main responsibilities, he replied:

> I don't know if it was my responsibility. I never had clearly defined responsibilities at BART... Responsibility must mean that I was responsible for something, that I could be -- if it were wrong, I would be called upon to -- according to that responsibility, to tell why I permitted this or that, this to be -- you have to have an authority in order to have a responsibility, is really what I want to say... I did not have an authorization. I did not have authority in that field.

But if Hjortsvang lacked the organizational responsibility for overseeing Westinghouse's activities, he clearly felt it was his personal and professional responsibility: "As you know, I considered this (following the development of the train control system) my duty. It was my personal duty."

It is hard to tell whether or not, during the first two years with BART, Hjortsvang's superiors were aware of, and approved of, his self-defined responsibility of keeping track of Westinghouse's development of the train control system. They did, however, approve his request to accept the Westinghouse invitation for BART engineers to go to Pittsburgh and work with the Westinghouse software design group. Hjortsvang went to Pittsburgh for ten months in 1969-70 to work with the system's designers in their Computer Systems Division.

At the end of his second year of employment with BART, Hjortsvang received a performance appraisal which was prepared by his supervisor E. F. Wargin. His overall evaluation was "Competent/Average -- A competent employee who performs satisfactorily the work expected of an individual of his experience." The strongest aspect of Hjortsvang's performance was summarized by his supervisor as follows:

> Broad knowledge and experience in electronics field and analytical ability at technical problems. Keen interest in his work and in the project in general. Willingness and ability to study, learn and take on new projects even at expense of personal affairs.

The weakest aspect of Hjortsvang's performance was summarized as follows:

> Slightly over-emotional and excitable. Occasionally this can blunt his logic. Does not always see the broad picture. Sometimes overmeticulous in technical matters but ignores detail in ordinary matters on the job.

Additional comments from the overall performance evaluation of Hjortsvang contained the following remarks.

> Has expressed a desire to specialize in the computer area of the BART-D System. He has the technical competence to do a good job on this but will require some assistance to develop an ability to supervise and work with others more easily.

The performance appraisal prepared on Hjortsvang at the end of his third year did not change substantially. He was again rated a "Competent/Average" employee with a good technical grasp and willingness to study new subjects. Weaknesses were described thus: "Sometimes avoids practical details. Somewhat set in his ideas, resists changes in routine."

Under a section labelled "Plans for Improvement" is a statement by Wargin that the employee (Hjortsvang) will undertake "an on-job study of computer facility and programming in conjunction with contractors personnel."

After his first two years with BART, Hjortsvang was assigned to the Westinghouse facility in Pittsburgh to learn more about the train control system. He went to Pittsburgh in September of 1969 and remained there for ten months. Hjortsvang reports that it was during this period at the Westinghouse facility in Pittsburgh that he first became concerned about the train control system and the general organizational problems at BART in trying to monitor the activities of PBTB and Westinghouse. He tells what first disturbed him about the Westinghouse operation:

> First of all, the organization within the Westinghouse Corporation, the job organization, there was two distinct divisions of Westinghouse. One is the Transportation Division, and another is the Computer Systems Division, and it was arranged that way, that the Transportation Division had sort of contracted the Computer Systems Division to do a certain job for a certain amount of money.
>
> And that was -- it looked like this interfered with the proper coordination of the efforts, of Westinghouse's effort, to develop this control system.

Hjortsvang's understanding of the relationships between the two Westinghouse divisions indicated that there was a great deal of inter-division conflict, and a consequent lack of adequate testing of certain system components.

"I was an observer. I was invited to meetings between people from the two Divisions and heard their argument against each other and disagreements about what should be done."

Hjortsvang felt that the organizational friction and conflict were reflected in Westinghouse's difficulty in fulfilling the contract requirements.

> The developments were going on in order to make components that would control the trains, and these components were, in my opinion, not adequate to do what the contract required them to do... I would say they were not properly tested or tried so that you could be sure that they would work the way they should. And the contract said that all elements in the train control system should be thoroughly tested for quality and reliability, and so on, and that was not being done... I can give you a example. We had the so-called antennas which were to pick up signals from track circuits from running on railroad tracks. These were things that later on proved to be very unreliable.

Although Hjortsvang was troubled by the inter-division conflict and the inadequate testing of system components, his primary concern was of a broader nature, involving the relationship among Westinghouse, PBTB, and BART.

> Remember, I was assigned to the Computer System Group that worked with the BART project, and what became apparent there after some months was that the...work was not being properly supervised. There were no weekly or biweekly meetings, progress meetings where the members of this group got together, presented their work, and the difficulties they had encountered, and where they were directed in such a way that the system would evolve as a total system and not a bunch of individual approaches.
>
> "And I know that, in the case of one program that had taken a person a whole year to make, that shortly after I left, had to throw the

> whole thing out and start it over again. And this was a direct result of this man never having been guided in his work, so his misinterpretation of the requirements just accumulated in time, and his efforts were completely wasted... And then I had another concern, and that was that PBTB, who were responsible for supervising the contract, came very rarely -- they sent one or two engineers over there perhaps once a month, and they had a meeting, and asked how things were going, and got some answers, but they never looked into matters themselves to see what actually was going on and how far they were. They just took some manager's word for it, and explanations, and I think at that time that if PBTB had done what I considered a proper job, had people on the premises to follow the daily work, and understand precisely what's done over there, the work, and the designs, try to evaluate the designs at that time, that we might have avoided a lot of the delays and other disappointments we have seen.

Although Hjortsvang's assignment to Westinghouse in Pittsburgh did not include making observations of how Westinghouse or PBTB was carrying out their part of the contracts, this is precisely what he took upon himself to do.

> The expressed purpose (of my assignment at Westinghouse) was that I should learn by doing, going over there, design some part of the system, and that I was to become very familiar with how it operated, and I was cautioned not to express any opinion or judgment to Westinghouse, because I was not entitled to do that.

But Hjortsvang did, in fact, prepare some five or six reports "telling about what I saw, and telling about the concerns I had," that were sent to his supervisor Mr. Andrews and to an engineer in the Transportation Division at BART. Hjortsvang indicates that the reports were acknowledged, but that there was "no response in reality. Nothing changed because of my reports."

When Hjortsvang returned to BART after his ten-month assignment with Westinghouse, he continued working on programs and learning as much as he could about the software portion of the system developed by Westinghouse. His office was moved from San Francisco to Oakland where some of the product of the Westinghouse system was being located, and which served as the communications control center. He continued to do much the same work he did in Pittsburgh, but he felt that his supervisors were not paying enough attention to his work and to the concerns he continued to have about Westinghouse and PBTB. Hjortsvang has stated:

> First of all, Mr. Wargin seemed to agree with what I had to say, but said, "That's nothing. There's nothing I can do about it. BART's policy is that this is PBTB's business and not our business. Let's not rock the boat."

Hjortsvang wrote several memos that were critical of PBTB, and suggested that BART should exercise its right to supervise them, since he believed that some of PBTB's decisions would affect the future adequacy or safety of the train control system.

The impact of Hjortsvang's reports and memos, issued while he was at Westinghouse and immediately after his return to BART, seems to have been nil. A large part of his work had been devoted to activities that he felt were the responsibility of PBTB. As a consequence, his work seemed to exist in limbo; it was not related to what BART was supposed to be doing and it was not part of his proper function, as his superiors repeatedly told him. Yet he nonetheless continued to do it because, as he stated: "I felt that these things were important to BART, and I was anxious to have this understood, have people look at this in a professional way, not in a political way."

Robert Bruder

When Robert Bruder joined BART, Hjortsvang had already completed three years of employment, including the assignment with Westinghouse and the early expression of concern about the actions of the contractors and BART management in relation to the Train Control System. Bruder's initial assignment involved coordinating train control and communication contracts. This meant that he maintained contact with the operations group in BART and with outside contractors, especially PBTB and Westinghouse. He has described his duties as follows:

> See that the contract was being complied with, monitoring the schedules and making recommendations and analysis of changes due to contract, presenting them through Mr. Fendel for Board, (of Directors at BART) approval when necessary... I would take the information (recommendations for change) and analyze it and discuss it with them (contractors) and understand it, and then either recommend it or not recommend it, and do some analysis of the cost and essentially write a summary page of what the change is about. And if they are over certain (amounts) -- $200,000 or over 10 percent of the contract -- they would have to go up through Fendel, Tillman, up to the Board for their approval, otherwise they would just file copies of the analysis and what the change involved was to cost.

Bruder states that he did not have a good working relationship with his supervisor Mr. Fendel. Why? "Because he didn't understand the technology that I was monitoring. I really couldn't communicate with him." Fendel was, nonetheless, generally accessible to people working under him. There was a meeting each Friday when the construction engineering group met to share information and discuss particular problems that might be coming up. People would take turns explaining how their contracts and their phases of the work were going. Fendel would take notes at these weekly meetings which he used for reporting purposes at this staff meetings.

> He (Fendel) more or less ran those weekly meetings. And it was his way instead of a bunch of written reports. Then on a monthly basis we'd

> write some summary report on general status and on schedule and major problems. The thing was very small, but those again would be summaries that went through Fendel to somebody else. I imagine an internal status report.

Aside from those meetings, Bruder and others worked independently, and there apparently was considerable satisfaction with these general working arrangements.

> We pretty well were left to our own...you could go out on the line someplace and just get lost... Everybody knew his field and he felt confident, and there was freedom within the group to go back and forth and ask questions if something came up and you needed support.

Yet despite this favorable working climate, Bruder felt that Fendel's limited knowledge of the technology resulted in inaction on many recommendations that Bruder put forward. He had on many occasions, including the Friday meetings, urged Fendel to do something about the lack of test scheduling by Westinghouse and PBTB. But Fendel did not respond to Bruder's urgings because he felt such matters were not the responsibility of his department. Bruder states:

> Well, if I really tried to force some action or follow through, it soon became apparent that, internally in BART, that we had no responsibilities on the train control in our group. The operations people had the technical responsibilities on those particular contracts.

Bruder characterized Fendel's reactions to his concerns: "Hey, thank God it's not in our group, its downstairs or in operations."

Bruder also expressed his concern about this problem to other engineers in the operations department and to consultants for PBTB, and he feels that they generally agreed with his assessment of the issue. He says that he proposed "that we get some kind of realistic scheduling out of the contractor (Westinghouse) and follow it through, starting with the details of the tests to be run, make the procedures more realistic, and some action of the work and design changes, and everything else that had to be done."

Even after meetings with fellow engineers which reached broad agreement on what the problem was and how it would be remedied, Bruder still felt that nothing was happening. "We were limited again by -- even if we all agreed on something -- by working through the PBTB engineers, and they in turn had to go to the contractor directly."

When Max Blankenzee joined the BART organization, Bruder had already been on board for 19 months and Hjortsvang for almost 5 years. Blankenzee's description of his early experiences with the BART organization seem to stress some of the same factors mentioned by Hjortsvang and Bruder -- the autonomy of the engineers to determine their own work activities and the absence of specific job descriptions and assignments.

> What was different was the fact that BART, its headquarters was in downtown San Francisco. The first installation for the computer system was in the basement by the Lake Merrit Station (in Oakland) where they built the control system and everything else in the central control area, and on top of that, the new building would have been built and is standing there now. And what happened is that we, as a group, Hjortsvang particularly and myself and the Westinghouse group were all located at the Lake Merrit Station in that basement. BART itself was over in San Francisco...when I came to work I went directly and worked within the basement...I spent most of my time away from the, what would you say, the home office, or at the time, the headquarters of BART. I was never introduced to anyone of the, what would you say, upper management that was above Ed Wargin. And I got a feeling that we had to stay as a standoffish mode.

Although he was hired by Ed Wargin, who was his supervisor, there was very little contact between them. Wargin worked in the headquarters,

> So there was very little relations, very little control and there was nothing like you have a job description, or say this is what we expect of you (or) well, Max, what you have to do, sometimes you might not be busy and sometimes you might, but you gonna have to find your own work... At that time there was no documentation, there was nothing really done on the total system. What I thought we would do is try to question Westinghouse and say, 'What have you done and how can I continue on to document and learn the system that you have designed?

Blankenzee describes his reaction to being located away from the main BART headquarters and his experiences at the weekly (sometimes bi-weekly) meetings which took place at the home office:

> Physically we were separate and I think it also made us feel that we were, in a way, somewhat of a bastard child. I use the expression widely really, because we did have to go to staff meetings (at the home office).. The staff meeting of the home office was more or less a review meeting by Mr. Wargin with PBTB, which was the contractor at the time, stating, hey, where are we now and what are we doing now? This type of a thing... Several of these meetings we would go to Wargin and say 'Ed, look, we got problems. We really got problems because PBTB really don't know what we're doing. At that time they really didn't have a guy that knew too much of what the system was all about, specially within the central control, the process, how to control it, how to measure it.

Blankenzee's description of the amount of "free" time he had, and his total discretion as to how it might be used, echoes earlier comments by Hjortsvang and Bruder:

> There was never a report asked, the only report that one time was asked to show how we could develop a maintenance documentation for the technicians on the computer system... There

> was nothing like that here, nothing controlled. And so... I was on my own. If I wanted to just help Westinghouse with the coding, I could have done that. If I wanted to lay around and read the books, I could have done the same thing... So complete freedom there which... I think if BART would have had a little bit better controlled environment, they probably not only would have seen these things we've talked about, but they would have stopped us, would have stopped us long before we ever got started on it. But because of our freedom that we had we would be able to look at the things the way we wanted to take a look at it -- be able to get involved in the things we wanted to get involved in.

Max Blankenzee and Holger Hjortsvang worked together at the Lake Merrit Station facility, but they were not close friends nor did they have a social relationship. In contrast to the professional tie between these two, was the relationship which Hjortsvang had with Robert Bruder. The two had know each other since Bruder joined BART in 1969. They lived in the same area and, at times, rode together to work in the same car pool.

The employment experiences of the three engineers before they joined BART is marked by considerable instability. All three experienced frequent job changes, because of lay-offs, dissatisfaction with their employment situation, or a search for better opportunities. The treatment they received from some of their employers is not that associated with professional employment. They were publicly criticized, "yelled at" before co-workers, and laid off with short notice. Though all three thought of themselves as professionals, some of the treatment they received from employers is hardly consistent with the label "professional."

After this somewhat unstable and demeaning record of early employment, it is understandable how and why each of the engineers would be greatly attracted to the BART position. For the first time in their careers they had a chance for stable, long-term employment, on a project that was being hailed as America's first effort into space age mass transportation. At BART, they could operate as truly professional engineers, on a project they could identify with because of its reputation for advanced engineering and its widely advertised benefits to a large public.

Once with the BART organization, however, the three engineers began to experience a diffuse authority structure, unclear task assignments, and relations with superiors which alternated between support and criticism. Their oscillation between routine monitoring of the activities of the BART contractors and excessive preoccupation with the failure of BART to maintain proper control of the contractors created confusion in their minds about their proper roles and is undoubtedly partly responsible for the problems they experienced. Their inability to get clarification or definition or -- at times -- even a response from their supervisors compounded the difficulty.

The three engineers were also able to expand their concern well beyond a narrow specialization or fragmentation. As a result, they were able to keep the "big picture" before them as they evaluated the progress of the contractors on various components of the system. The mere fact that they were so concerned about being behind schedule, which should more properly have been concern of management, indicates the degree to which they identified personally with the goals of BART. Professionals limited in their tasks to narrow specializations would have been less likely to "see" the things observed by Hjortsvang, Bruder, and Blankenzee.

What each of these men learned about the entire BART organization during their early period of employment had an important bearing on what they perceived as problems with BART management ...and with the development of the automatic train control system. Much of their behavior between May of 1971 and March 3, 1972 was an outgrowth of these perceptions.

Chapter 5

Protecting the Public Interest:

The Role of the Technical/Professional Societies

Technical and professional societies abound in American society. They reflect the development of occupational specialization and of technical and professional programs in higher education. Such occupationally based associations are faced with persistent problems concerning both their general goals and specific activities. One problem concerns the relative emphasis the association gives to two different goals: (1) improving and protecting the economic status, prestige, and working conditions of the professional; and (2) providing the means to facilitate the development and communication of new knowledge relevant to professional activities. A second problem concerns questions about how and to what degree the societies should become politically involved by committing the society to be in support of or opposition to, specific public issues.

Questions of this nature persist in professional/technical societies because members have differing backgrounds and interests, and differing views of the proper role of their societies. Membership diversity may be based upon educational experience, field of specialization, or employment setting. One can find in the same professional/technical society managers of corporations, professional employees, academics and entrepreneurs.

Such diversity of member interests makes it very difficult for professional/technical societies to "speak with one voice" on a great many matters, including the complex issue of professional ethics.

The September 23, 1969 meeting of the Contra Costa County Board of Supervisors droned on. Slowly the Board worked its way through its agenda; the few people in the audience waited patiently for the item which had brought them to the meeting. Finally, the time came. The Board of Supervisors was ready to fulfill its statutory obligation of appointing two of Contra Costa County's four representatives to the Board of Directors of the Bay Area Rapid Transit District. As this agenda item was called, two men in the audience, Gilbert Verdugo and Roy Anderson, straightened up.

The terms of office of two of the BART Board members appointed by the Contra Costa County Board of Supervisors would expire on October 21. One of the incumbents, H. L. "Jack" Cummings of Martinez, age 81, had indicated a desire to retire from the Board. The other incumbent, Joseph S. Silva of Brentwood, age 70, clearly wished to be reappointed for another term.

Silva had served eight years as a County Supervisor. Although he liked to describe himself only as a "poor Portuguese farmer," Silva was a man of some influence in the County. He had served terms as mayor of Brentwood, as city councilman, justice of the peace, and school board trustee. In addition, Silva had a personal commitment to BART. It was his vote, as a member of the Contra Costa County Board of Supervisors in 1962, that broke a deadlock on the issue of BART and paved the way for the approval of the rapid transit system. The Alameda County and San Francisco supervisors had already approved the plan, but Contra Costa County Board members were stalemated at a 2-2 vote. After Silva finally voted "yes", the issue went to the voters in the three counties, who approved BART by a narrow margin.

Four days prior to the September County Board meeting, on September 19, 1969, Roy Anderson had become the sixth announced candidate for appointment to the BART Board. The other candidates for the appointment, along with Silva, were State Assembly Speaker Luther

Lincoln of Concord, Lafayette attorney Harold Mutnick, Richmond businessman Nello Bianco and Lafayette industrialist John R. Lavinder. Anderson, a registered professional engineer, was a bridge construction specialist with the State Department of Public Works and was a member of the liaison engineering staff on BART's TransBay Tube. Although there were no other engineers on the BART Board, it was not Anderson's profession which made him an unusual candidate. Rather, it was the fact that he was a candidate with a sponsor. Roy Anderson was the nominee of the Diablo Chapter of the California Society of Professional Engineers. If the nomination of a candidate for a seat on a public board by a chapter of a professional society was not unique, it was, at the very least, highly unusual. Typically, professional engineering societies have busied themselves with professional aspects of engineering practice, not with the more political issues of selecting board members of public bodies.

Gil Verdugo, a fellow member of the Diablo Chapter, was the person who was to place Anderson's name in nomination. When the Supervisors indicated readiness to receive the names of candidates for the BART director's seat, Verdugo rose.

> Gentlemen I came before you tonight to introduce to you a man who would make an excellent representative of Contra Costa County on the BART Board of Directors. All of us in Contra Costa County are excited about the possibilities of the growth and development that the BART system will bring to our county. I have worked for the division of highways and I know the shortcomings of the complete automobile-oriented transportation system. Therefore, I have long been a supporter of a multimodel system of transportation. We need BART.
>
> The Diablo Chapter of the California Society of Professional Engineers supports BART and is anxious to have the system brought into revenue service as quickly and efficiently as possible. In the spirit of helping to achieve this end, the Diablo Chapter nominates and recommends for

> your consideration Roy Anderson as a perspective appointed member to the BART Board of Directors representing the people of Contra Costa County.
>
> Roy Anderson has a Master of Science in civil engineering with a specialty in transportation and he is a man who already knows much about the BART system. He is currently a traffic safety and control engineer in the division of Bay Toll Crossings in the state of California transportation department. From April 1966 to September 1969, he was the principal assistant to the liaison engineering on the BART Transbay Tube and Approaches. During this two and a half year period, he participated actively in the formulation of policy procedures and their application for monitoring all phases of administration of contracts funded by $180 million of state toll revenues. He was in continuous contact with the BART staff and with their consulting engineers to assure conformance with all agreements, to assure construction in accordance with plans and specifications, to review and approve all changes and to approve payment for contract work performed.
>
> From September 1968 to July of this year 1969, he was on a leave of absence from the division of Bay Toll Crossings in order to attend the University of California at Berkeley. At the University, Mr. Anderson studied transportation policy and planning and was awarded the master of science in civil engineering degree. In the few months since the receipt of his masters degree, Mr. Anderson has been a traffic safety and control engineer in the division of Bay Toll Crossings.
>
> After you hear from Roy Anderson, I'm sure you will agree with us in the California Society of Professional Engineers that Roy Anderson would make an excellent Director of BART.

Gil Verdugo settled back into his chair.

The Supervisors, unused to formal institutional nominations for positions on the BART Board, thanked Verdugo. Had they been more conversant than they were with the structure and functioning of professional and technical engineering societies, they would have been even more surprised than they were at the action of the Diablo Chapter in proposing a candidate for the BART Board.

Professional and Technical Societies

Professional societies, as the name implies, are those associations of engineers which tend to be concerned with the "professional" aspects of the engineering career. Typically, the professional society is concerned with the engineer's working conditions, with his salary and with the character of his professional behavior, as a guide to which the society will frequently promulgate a code of professional ethics. The technical society, on the other hand, tends to concern itself with the technical aspects of the profession: the development of new knowledge and technique and the application of these in the solution of typical engineering problems. Neither the professional nor the technical society, however, has ever played a significant political role--the role that the Diablo Chapter was moving into by proposing Roy Anderson as its candidate for the BART Board.

Both Gil Verdugo and Roy Anderson were civil engineers and, as such, had a particular interest in a basically civil engineering venture of the scope of BART. And it is civil engineers who tend, more than those in any other specialty, both to be registered professional engineers and to be active in professional societies. Now, who are the engineers listed as "registered professionals"?

Each state and organized territory of the United States has a law requiring engineers to be registered in order to practice in the profession. In each one of these governmental units save one (Montana), the engineering registration law includes an "industrial exemption clause". This clause exempts those engineers engaged in the design or fabrication of manufactured products from the requirement to be registered. As a consequence, the rolls of registered engineers contain a predominance of engineers engaged in nonmanufacturing activities, which means, typically, a great many civil engineers and relatively fewer from the branches of engineering generally involved in manufacturing: chemical, electrical, mechanical, aeronautical, industrial and so forth. Now, although the registration of an individual engineer to practice engineering within a given state or territory is completely independent of any membership that that engineer may hold in any professional or technical society, a greater than expected share of members of the professional societies tend to be drawn from civil engineering. The "professional" designation of their registration appears to carry over to their self-conception and to lead to participation in professional society activities. Both Gil Verdugo and Roy Anderson were civil engineers, were registered professional engineers, and were active members of the California Society of Professional Engineers.

The California Society of Professional Engineers is one of more than 50 such societies, one for each state and territory. The state chapters are composed of a number of local chapters. The Diablo Chapter which nominated Roy Anderson for membership on the BART Board is one of 35 such local chapters. The state society is a confederation of the local chapters and the National Society of Professional Engineers (NSPE)--the national professional engineering organization--is a confederation of the State Societies. Unlike the professional societies, the technical societies of the engineering profession are typically organized, not by geography but rather on the basis of the subdisciplines of the profession, e.g. The American Society of Civil Engineers, etc. The technical society most deeply involved with the BART incident is the Institute of Electrical and Electronic Engineers

(I.E.E.E.) which will be discussed in Chapter 13.

Unlike the technical societies, which are relatively highly centralized, the professional society is primarily and fundamentally a "grassroots" society. That is to say, the local Chapters of the State Society have considerable autonomy. Article V, Paragraph 2 of the California Society of Professional Engineers (CSPE) Constitution says, "In all matters of local concern not covered by the Society's Constitution and Bylaws, chapter or student chapters shall retain full autonomy, but may call upon the Society for advice, counsel and assistance." At the state level a different kind of subdivision within the organization is provided by having separate groups associated with professional engineers in industry, professional engineers in private practice, and professional engineers in government.

The Preamble to the Constitution of the CSPE describes its basic goals;

> Recognizing that service to humanity, to our nation and to our state is the fundamental purpose of professional engineering, the California Society of Professional Engineers does hereby dedicate itself to the promotion and the protection of the profession and of engineering as a social and an economic influence vital to the welfare of the community and of all mankind.

The objectives of the Society "shall be the advancement of the public welfare and the promotion of the professional, social and economic interests of the Professional Engineer." (Article II).

The CSPE, like the other state societies and their local chapters, and like the National Society of Professional Engineers has traditionally placed great emphasis on defining and monitoring proper ethical practice. The California Society has an Ethical Practices Committee and the National Society has a Board of

Ethical Review. NSPE has been and continues to be a vigorous promulgator of their Code of Ethics for Engineers (see Appendix). Because of this concern with issues of engineering ethics, it was not unusual to have members of the CSPE actively interested in the affairs of BART. Several years before the September 1969 meeting of the Contra Costa Board of Supervisors, several members of CSPE, particularly a group from the Golden Gate Chapter, took an active and public interest in the contract between BART and the engineering joint venture, PBTB. Most of the interest had been expressed by members of the three chapters whose localities would be most affected by the BART system: the Golden Gate Chapter, comprising essentially San Francisco and South San Francisco, the East Bay Chapter which is basically Oakland and Berkeley, and the Diablo Chapter which is the area roughly the same as of Contra Costa County.

BART, PBTB and CSPE

The relationship between BART and PBTB began on May 14, 1959, when BART retained the services of the joint engineering venture composed of Parsons, Brinckerhoff, Hall and MacDonald, Tudor Engineering and the Bechtel Corporation to develop a regional rail plan. Following voter approval of the BART bond issue BART signed another contract on November 29, 1962, with the engineering joint venture firm composed of Parsons, Brinckerhoff, Quade and Douglas (the successor to Parsons, Brinckerhoff, Hall and MacDonald), Tudor and Bechtel. This firm, called PBTB (Parsons, Brinckerhoff, Tudor and Bechtel) contracted to direct the total engineering task of BART from overall system planning through research and development, design, and management of construction. Even before this contract was signed, however, its terms became known among the engineering community in the San Francisco area, and many engineers spoke out publicly concerning the high profit to the joint engineering venture which appeared to be guaranteed by the contract. In addition, they began to express their dissatisfaction with the contract to the Ethical Practices Committee of the California Society of Professional Engineers.

These concerns were basically of two kinds. First, engineers were disturbed that only the PBTB consortium had had any realistic opportunity to obtain the contract. This is not, however, to suggest that there was any advocacy of opening up the contract to bids. Indeed, within the practice of engineering, it is not considered ethical to bid for engineering services, since it is consistently argued that engineering services should be performed, not by a low bidder, but by those persons and organizations which are professionally best qualified to do the job. Thus, no obvious breach of engineering ethics or engineering practices was present when BART chose their engineering consulting firm without going through a bidding process (as is usually required in construction contracts). Still, an organization looking for engineering services is generally advised to have at least some preliminary discussion with a number of firms before selecting the firm to do the job--and this is precisely what BART did not do before entering into its contract with PBTB.

The second and ethically more substantive area of concern expressed to the Ethical Practices Committee was simply that the terms of the contract seemed inordinately lucrative to the engineering consultants. The key provision was that the consulting engineering organization had to absorb essentially none of the overhead costs of the project. For example, it is usual and expected that an engineering consulting organization would take its secretarial, its internal accounting, and its top level management expenses out of its "profit", charging those items as an overhead burden to be internally absorbed by the organization. This was not the case with the contract between BART and PBTB. Rather, this contract called for the three firms to be reimbursed by BART for salaries paid to virtually all employees working directly on the project and for all major cost items. In addition, they were paid 1-1/4 time the payroll for central office employees, 9/10 of the payroll for field office personnel and 1/10 of subcontracting, technical consultant and architectural costs.

In the face of wide-spread criticism of the BART-PBTB contract, John M. Peirce, first General Manager of BART and a former State Director of Finance, and Kenneth M. Hoover, Chief Engineer for BART, defended the contract on the grounds that it provided that all engineering subcontracts would be subject to BART Board Approval, and that the three engineering firms would submit salary ranges for classifications in advance. Peirce insisted publicly that

> the district will have strict budgetry control in the same manner used by government agencies generally. Our district controller has accountants and budget experts. We will pre-audit all expenditures by the engineers...The procedure spelled out in the contract is in strict conformity with the best governmental fiscal practices...This will be no "blank check" in any sense of the word.

Peirce and other spokesmen for PBTB and for BART further defended the contract's fee formula as conforming to a standard formula within the industry. Moreover, they emphasized that since the BART contract was unique in character and would probably be the largest contract ever let for engineering services, special provisions to meet the project's needs were justified.

Most of the concerns expressed to the Ethical Practices Committee of the CSPE came from members of the Golden Gate Chapter in the San Francisco area and from Oakland-Berkeley area engineers who were members of the East Bay Chapter. A number of discussions were held and at least two meetings took place among representatives of BART, PBTB, and CSPE. It appears that, after a good deal of public discussion, a closed hearing was held to consider the ethical appropriateness of the terms of the contract. The CSPE representatives were apparently not impressed by the arguments they heard, since following this hearing, a public statement was made by CSPE to the effect that the terms of the contract were inappropriate and exorbitantly expensive to the public.

Typical of comments made by CSPE members active in this investigation are the following: "The BART people were complete neophytes at contracting."

"The BART contract is the greatest rip-off I've ever seen."

"Bechtel had the BART directors in their pocket."

In addition, CSPE members active in this investigation report that a variety of pressures were put upon them to cease and desist in their public investigation and discussion of the terms of the contract. The President of the NSPE is reputed to have called a CSPE member active in the investigation to tell him to back off from his activity because he was damaging the reputations of nationally prominent engineering firms. A representative of Bechtel is quoted as saying to another CSPE activist, "There's enough work in this for everyone. Don't make waves, or you won't see a nickel of the subcontracting."

A Question of Professional Ethics

Because of the intensity of feeling involved and the national visibility of the BART Project, the actions of the CSPE members began to attract the attention of the national organizations. Traditionally, when situations of questionable ethical practice are brought to the attention of the NSPE Board of Ethical Review, the response is to deal with them on a general and abstract level. This is typically done by disguising the real situation and creating an abstract model of the ethical issue. The ethical questions, presumably unaltered by the disguise, are then discussed in meetings of a Board of Ethical Review specifically chosen to consider the case. The Board considers the case from the point of view of the appropriate sections of the Code of Ethics and an "opinion" is prepared. Periodically the opinions of the Board of Ethical Review are collected and published by NSPE.

The Opinions of the Board of Ethical Review, Vol. 2, 1967, contains case No. 66-1, Engineer's Criticism of Fees. The abstract set of facts in this case are reported as follows:

> A local public body signed a contract with an engineering firm for complete engineering services for a new airport, including the establishment of fees for preliminary planning, general consulting services, preparation of construction plans and specifications, field engineering during construction and other technical services, including coordination of a unique mechanical passenger conveyor system with a basic design.
>
> Certain public officials charged publicly that the fee structure was excessive and the question was referred to a grand jury. The controversy received considerable publicity through the local press and on radio and television stations. During the period of public discussion of the fee structure, a group of local consulting engineers, none of whom who had airport design experience issues a report, concluding that the fee was substantially in excess of the fee schedule published by the state professional engineering society. The report of the local group was made public and received general press and radio and television coverage.
>
> QUESTION: Is it ethical for a group of consulting engineers to issue a public report criticizing the fee arrangements contained in a contract with an engineering firm?
>
> The Opinion case as published makes reference to the Code of Ethics

Section 2(b) -

He shall seek opportunities to be of constructive service in civic affairs and work for the advancement of the safety, health, and well-being of his community.

Section 4(a) -

He shall not issue statements, criticism, or arguments on matters connected with public policy, which are inspired or paid for by private interests, unless he indicates on whose behalf he is making the statement.

Section 5 -

The Engineer will express an opinion of an engineering subject only when founded on adequate knowledge and honest conviction.

Section 5(a) -

The Engineer will not attempt to injure, maliciously or falsely, directly or indirectly, the professional reputation, prospects or practice of another engineer, nor will he indiscriminately criticize another engineer's work in public. If he has knowledge that another engineer is guilty of unethical or illegal practice, he shall present such information to the proper authority for action.

The discussion as presented in the case ultimately keys on the phrase, "none of whom had airport design experience" from the facts section, and on the phase "adequate knowledge" in Section 5 of the Code of Ethics. The Opinion concludes that it is <u>not</u> ethical for a group of consulting engineers to issue a public report criticizing the fee arrangements contained in a contract with an engineering firm under the circumstances described.

Interestingly, the ethical question addressed involves only the behavior of the critics of the BART-PBTB contract, and not the ethical characteristics of the Contract itself.

We see in this first contact between BART and the California Society of Professional Engineers a preview of the relationship which would be established again in 1972 when engineers Hjortsvang, Bruder and Blankenzee were fired. Specifically, we see individual members of the California Society of Professional Engineers becoming concerned both as engineers and as citizens, over something going on which is internal to the BART/PBTB operations. These members raised their concerns within the appropriate channels of the California Society of Professional Engineers and they helped to draw public attention to the issue. A reaction then developed both within and outside the Society to bring pressure to bear on the object of this public criticism. Finally, both in 1962 and again in 1972, individual members who were particularly outspoken in their views suffered professionally because of their activity.

Roy Anderson Speaks

In the context of this view of the character and role of the professional society, let us return to the Contra Costa County Board of Supervisor's meeting on September 1, 1969. Following the nomination of the various candidates for the appointment to the BART Board, the Supervisors asked for further comment. After a pause, Roy Anderson, the candidate of the Diablo Chapter of the CSPE rose to his feet. He began:

> Mr. Chairman and Members of the Board: The need for adequate transportation in a society as complex as ours is questioned by no one. The matter of what kind, how much and how financed is hotly debated by many. One thing is clear: the approach in California of applying rural transportation solutions to an urban setting can only bring disaster to ourselves and our environment. Freeways to transport commuters is not the answer. We can no longer cover our landscape with asphalt and concrete and spew deadly poison into our atmosphere.

The voters of the Bay Area Rapid Transit District most obviously recognized the pressing need for a solution in 1962, and, with generosity and faith in their government, they encumbered their homes and businesses in the hope that a solution had been found. They were promised a ride in 1969 on the finest and most modern rail system in the world. In 1969, still years away from completion, the taxpayer has indeed been taken for a ride, but not on a train. The taxpayer has been faced with the decision of whether to pay more so he can have the ride he was promised or deny funds and go nowhere. He chose to pay. This attitude was no doubt expected by the consulting engineers that prepared the feasibility report.

The question of how the deficits grew has not been answered factually, but has been covered over with layers of rhetoric. The fragmented method of selecting Board Members and the failure of our elected officials to make changes in the Board representation and to demand a thorough audit by the State is both perplexing and mystifying.

The stockholders in a private corporation would certainly demand that management changes be made if deficits occurred year after year. The BART-D Board only recently, after new taxes had been added to fund the reported deficits, and with almost unbelievable disdain for the taxpayer, raised the salary of the executives to an incredible level under the circumstances. If the Board rewards failure so freely, then what, may I ask, can we expect in the future. We are already told they are broke again.

What the public can expect in the future under the present leadership, is more bad management and deficits in operation of the system. BART-D will be back for more money to operate if reform is not made now. The BART-D management is like a cracked record - BROKE AGAIN, BROKE AGAIN, BROKE AGAIN...

What are some of the reasons that BART-D is in trouble?

The Board, consisting of persons not knowledgeable in engineering and public administration have relied almost totally on their general manager and staff to guide the Board decisions. This is not unusual. The Board's great failure has been in the apparent poor choice of general manager and the resulting staff selected by the general manager.

The BART-D Chief Engineer in 1962 recommended the award of the engineering contract to the same organization that made the feasibility report. The contract was without meaningful controls as is evident today with fees increased from $47 million in the original contract to an estimated $89 million plus approximately $13 million for the TransBay Tube and Approachs. The contract should not be too surprising when one considers that the BART-D chief engineer had been an employee of the Consultant a few years earlier and was recommended to the BART post by a principal in the consulting organization. The public interest was not guarded.

I have observed that the BART-D engineering staff has an apparent attitude that the consulting engineers are in charge and should not be challenged in their decisions. The attitude is one of "don't rock the boat." Consequently, the consultants do in fact run BART-D, by default.

The Board's appointment of a general manager who lacks knowledge and experience in management, engineering and contracting has no doubt been the most costly mistake the Board of Directors has made and they still refuse to make a much needed change.

Another hiring practice that seems questionable was the reported hiring of a member of the Legislative Analyst's staff who reviewed and approved the validity of the need and adequacy of the $146 million requested by BART-D. The Legislature passed the tax bill and now BART-D declares those funds inadequate. This type of action certainly leaves the District open to criticism.

All of the events mentioned plus many others that have taken place have created a credibility gap that the present Board of Directors and staff cannot overcome. Because of the failure of the local and state officials to bring about the needed change, there has been created a mistrust of the local government's ability to manage the taxpayers' hard earned dollars entrusted to them. The repercussions come through the taxpayer voting against bonds for schools, juvenile halls, jails and other such proposals. The voter has few avenues of regress (sic) against independent boards such as BART-D when the elected officials refuse to act responsibly.

To this Board of Supervisors and to the citizens of the District I would like to state that positive action must be taken now if we are to bring reform to BART-D. The Board has the power here today to appoint two new Directors and charge them with the responsibility to doing all within their power to bring reorganization and responsibility to BART-D.

The following are a few suggestions I will make to the approach that should be taken:

1. Dismiss the present General Manager.

2. Retain a top national management consultant to recommend names to the Board of Directors for the General Manager position.

3. Retain a management consultant to study organization and make recommendations.

4. Roll back the latest salary increase of staff executives.

5. Request complete audit by the State.

6. Hire an adequate engineering staff to assure Directors that all contracts are administered properly and to review all matters requested by the Board.

7. Make a complete report to the public concerning the true status of BART-D utilizing

> the State audit and the management consultants report. State the priorities and plans for completion and operation of the system.

With the aforementioned items, I believe the District can begin to function in a business-like and responsible manner that can bring order to the chaos that has preceded and regain the support of the public. Government is created to serve the public interest and not itself - that fact has been overlooked by BART-D in the past.

My reason for accepting the Diablo Chapter of C.S.P.E. nomination for the Directorship was based on my concern for efficient and responsive government. My experience in dealing with the District staff and their consultants in the administration of their contracts has left me in grave doubt as to their ability to administer public funds.

The public interest has not been placed foremost. A thorough investigation by State auditors experienced in public works administration will reveal some shocking circumstances. A thorough investigation is necessary, and if the BART-D Board fails to request an independent investigation (which is doubtful without a change of the present majority) then the elected local and State representatives should so request.

Whether or not this Board of Supervisors appoints me to the BART-D Board of Directors, I urge you to express your desire for change in BART-D's past policies by appointing two new and qualified persons. The public deserves no less.

Thank you.

With that, Roy Anderson settled back in his chair. The Chairman of the Board of Supervisors thanked both Anderson and Gil Verdugo for their presentations and indicated that their recommendations would be taken under advisement by the Board of Supervisors.

The next morning, the Board met in Executive Session to make its appointments. The action it took was to reappoint Joseph Silva, "the poor Portuguese farmer" to one seat, and to the other, to name Nello Bianco, the Richmond businessman. The Oakland Tribune on September 24, 1969 reported that these choices were made in secret session. The Supervisors later revealed, however, that the vote for Silva was unanimous and that for Bianco, 4-1. After the Supervisors' appointments were announced, Bianco told the Oakland Tribune that he would critically review all of BART's operations, but he refused to comment on Anderson's charges. Bianco, 41 years old, owned the Capri Delicatessen and Catering in Richmond. He was also the chairman of the Richmond Personnel Board, was active in Democratic politics in the western end of the county, and is said to have had the strong support of supervisors from San Pablo, Richmond, and Concord.

The Wednesday, September 24, 1969 edition of The Independent reported in detail the suggestions for change made by Roy Anderson during his talk to the County Board of Supervisors. It further quoted Anderson as follows:

> (the BART Board) consisting of persons not knowledgeable in engineering and public administration, have relied almost totally on their general manager and staff to guide the Board decisions. This is not unusual. The Board's great failure has been in the apparent poor choice of General Managers and the resulting staff selected by the General Manager...I have observed that the consulting engineers are in charge and should not be challenged in their decisions.

Soon after this newspaper coverage of his presentation to the Supervisors, Roy Anderson was called into the Office of his Supervisor, the Chief Engineer of the Bay Toll Crossings Division. The Chief Engineer expressed concern about the publicity being given to Roy's feelings and attitudes toward BART.

> Roy, you know how hard we're working to get the South Bay Bridge idea accepted. We cannot afford even the appearance of competing with or being against the BART system. The publication of your views is putting this division in a very difficult position. As a public employee in this division, you have only second class citizenship rights; you do not have the ability or the capability to speak out on public issues. I do not want to see your name associated with any public criticism of the BART system.

For the next two and a half years, Roy Anderson maintained a public silence on the subject of BART. Only after Hjortsvang, Blankenzee and Bruder had been fired, did he once again begin to speak.

Part II
Crisis

Chapter 6

The Gathering Storm: The Engineers' View of the Clash Between Professional Responsibility and Organizational Authority

> Organizations are composed of persons with diverse backgrounds, goals, interests, and values. Such differences provide the basis for much of the conflict between individuals and units within organizations. On occasion, the differences become so pronounced that employees feel that extraordinary action is required to resolve problems they perceive. One possible action is simply to leave the organization. Another action is to try, by unorthodox means, to change the actions of others in order to bring the organization more in line with the individual's expectations.
>
> An employee's decision to oppose established practices and authority in an organization--to engage in whistle blowing-- is a serious step and typically does not occur in response to a single event or conflict. Rather, it is the culmination of a series of actions and reactions that have previously taken place in the course of his relationship with his employing organization. To fully understand whistle-blowing, it is necessary to trace the sequence of events that leads professionals to conclude that no other course of action is open to them.

By the time Max Blankenzee joined the BART organization in May of 1971, there already existed a climate of concern about the technical aspects of the train control system and the inability or reluctance of BART management and its engineering staff to monitor and direct the activities of the contractors, PBTB and Westinghouse.

Well before this time, Holger Hjortsvang and Robert Bruder had tried to get BART management to do something about a control system which they believed was not adequately documented or tested. They failed. But with the addition of Max Blankenzee to the BART organization, Hjortsvang and Bruder were moved to go "outside" the organization's authority structure to get what they regarded as positive action on their concerns.

Examination of the actions of the three engineers forces consideration of several important questions which face professional employees of large modern organizations.

1) What are the organizational conditions that lead an employee to resist organizational authority to a point of risking his own job and career?

2) Why does a professional employee decide he has not satisfied his ethical responsibilities by simply expressing his concerns to superiors? How and why does the demand for action on these concerns constitute an ethical responsibility?

3) What personal and professional background and attitude lead professionals to undertake actions which can result in their dismissal? Ethical commitment? Technical arrogance? Political naivete?

Learning to Oppose Organizational Authority

Analysis of political revolutions by social scientists suggests that most successful revolutions have been preceded by unsuccessful ones. It seems that people must learn how to revolt, and the abortive efforts apparently provide the necessary harsh learning experience for those who would be successful revolutionaries. From this conclusion we may analogize that, just as revolutionaries must learn how to organize their resources and revise strategies and tactics to overturn established authority, so individuals must learn to organize their own moral resources in support of a battle against authorities intent upon resisting them. If we may be permitted this analogy, it might be useful to view Hjortsvang, Bruder, and Blankenzee as people who had to learn how to oppose organizational authority. They had to define the problem that faced them, develop a commitment to their cause and a belief in the correctness of their purpose, and hold a view of their opposition that required them to commit their (from the organization's standpoint) irregular and illegitimate actions.

The early concerns of Hjortsvang and Bruder about the design of the train control system, the lack of adequate schedules for its testing, and the failure of BART management adequately to monitor the work of PBTB and Westinghouse persisted throughout 1970. While PBTB apparently made some efforts to get Westinghouse to establish realistic schedules for testing and delivery of the system, the engineers believed that such correctives were, at best, piecemeal in nature and were not really responsive to the general problems that they had been bringing to the attention of their supervisors.

Bruder has commented on why he was not satisfied with the steps taken by PBTB to get Westinghouse to set up a testing schedule.

> Well, the total schedule wasn't really laid out, so anybody concerned and involved in these contracts knew it was going to go on and the progress really made against the end schedule that a realistic contract would have required. Or to put it another way: The contractor didn't totally commit himself to try to hit a target. In other words, the manpower needed for the testing and the time left on the schedule and depending on who you talked to or how you interpreted the contract, BART had delayed the contractor to the point where he had some delays that were legitimate delays to his schedule.

Bruder has also indicated his belief that all levels of the BART organization did not share the same knowledge and understanding of the scheduling problem.

> the message wasn't getting up to the Board of Directors of where we actually were realistically, that you could responsibly tell the public when you were going to open. The two weren't in alignment. They were telling the public one thing and realistically inside we knew we couldn't make it.

Hjortsvang's chief concern was that the problems he had observed while he was assigned to Westinghouse in Pittsburgh were being dealt with on a piecemeal basis. This approach was making the technical problems progressively worse rather than better. In his words, "My realization that...the software system was in danger of never being satisfactorily completed, my suspicions there became more and more pronounced as I saw what was happening. The thing grew out of all proportions."

Hjortsvang has described Westinghouse's approval of developing the software system as very unsystematic. For example, the software was to provide corrective strategies for any deviations from the normal operation of trains that occurred. But Hjortsvang observes:

> I think my concern was that there was not a clearly expressed definition of the goals, the end purpose of these (corrective strategies). The way the job was being worked was that, "Let's look at a specific upset in the system, a train being late; let's write a program that will correct that," and then another person will look at another kind of difficulty, let's say a train that is stalled on the tracks, and what precautions should be taken in order to minimize the disturbance caused by a stalled train. There was a lack of what I would call systems engineering in the concepts.

Hjortsvang continued to express his concerns about the train control system to his supervisor, and to engineers he knew in the Westinghouse group. But he felt he was not getting a proper hearing from his supervisors, and that, for organizational reasons, he was limited in his ability to influence people at PBTB. For example, Hjortsvang was asked if he ever considered writing a memo to Mr. Quintin of PBTB, who had the major responsibility for the writing the specifications for the train control and communications system. He replied: "No, no. I have no business of writing Mr. Quintin. I could not, as a BART employee, write to an employee of PBTB." However, any of Hjortsvang's superiors could have communicated with Mr. Quintin: "Like Mr. Kramer. Even Mr. Wargin could have done it."

Hjortsvang did, however, express his concerns orally to Mr. Quintin during periodic meetings with PBTB personnel. It appears that Quintin never agreed with Hjortsvang during these meetings, but the issue was not so much his technical analysis as his right to be concerned with such matters.

> He didn't disagree with my analysis. He said that Westinghouse was responsible for it, and the functions were correct. That was all that PBTB could be concerned with. How they performed these functions were none of his concern...I didn't think from what I have seen, that Westinghouse was worthy of that confidence, that if we can just let them do what they want, as long as in the future they will come up with the functions that have been prescribed, and everything is all right.

After numerous unsuccessful efforts to get a hearing for his views from his immediate supervisor, Mr. Wargin, Hjortsvang began to send memos to BART management at levels above Wargin. Hjortsvang, asked if these memos indicated that he was "going outside normal channels," i.e. over the head of his supervisor, took exception to this interpretation of his memos.

> I did not keep Mr. Wargin in the dark about what I did. I may not have given him copies of memos I sent, for example, to Mr. Kramer or Mr. Ray, but I have--I discussed the problems many

> times with Mr. Wargin and he--I let him understand what I wanted to be known. So he knew that I got in touch with these other people.
>
> No, I wouldn't say that (my contacts with Mr. Kramer and Mr. Ray were outside the normal channel of communication.) I think it's quite normal if you discuss your problems, your concerns, with your direct supervisor and he doesn't--he doesn't take any action, that you will tell him that you are going to present this to other people in the same department.

When Hjortsvang felt that he did not receive a satisfactory response from Mr. Kramer to his memos on January 7, 1971, he sent a memo to Mr. E. J. Ray, Kramer's superior. For unknown reasons, this memo never reached Ray, according to what he later told Hjortsvang. The memo to Ray, included in its entirety, reads as follows:

> PROBLEM
>
> The yard interface problem, described in the attachment, serves as one example to demonstrate the necessity of technical, coordinating function within the BART organization. Although the example problem falls in the "start-up" category, other engineering and programming problems with impact on different divisions of the Operations Department, as well as on other departments or external entities, will continue to arise.
>
> As mentioned in my memorandum dated Dec. 2, 1970, it appears that Mr. Ray realized the need for coordination of rules and operating procedures when the office of the Start-Up Coordinator was established in January of 1970. To my knowledge, no such coordination has been attempted by Mr. Kramer, but the need for this has become more and more urgent to people concerned with system aspects. In particular, we are missing a channel through which BART's considered requirements and interpretations can be communicated to Westinghouse system engineers at the proper technical level.

The problems we face in this area are not unique to BART. In Bulletin 1658, 1970, the U.S. Department of Labor has reported on Computer Process Control and Manpower Implication in Process Industries. None of the installations surveyed are as complex as the BART system, but the organizational and manpower implications of computer control are well illustrated. I recommend that BART officials, concerned with system operation, read this publication to gain an understanding of, what a computer controlled system is, and what staffing requirements it implies.

ARGUMENT

During my time at the Computer Center at Lake Merritt the constant need for direct communication and technical decisions gave rise to much thinking about ways to improve the situation. These thoughts have gradually chrystallized into a concrete proposal.

Prior to this, I had attempted to gain support (unsuccessfully) for an Equipment Maintenance Section for Lake Merritt Central, under the Power and Way Division, with the proviso that the supervisor be authorized to communicate directly with other divisions and sections. Reference, "Policy Proposal" of Aug. 4, 1970.

Meanwhile, it became increasingly clear that equipment maintenance is a minor part of the job at hand. Central Control extends to all parts of train operation, and requires understanding leadership coupled with intimate system knowledge, to merge the efforts of those responsible for the different aspects of operation, into the Automatic Train Control scheme.

This broader view led to the concept of a new separate Division where all systems engineering, analysis, and programming could be concentrated, to constitute a service organization, in the same sense as the Data Processing Department.

Specifically, the new Division would be named the Systems and Programming Division of the Department of Operations.

In addition to the Systems Engineer, it would include a Train Control Engineer and his assistants, to deal with electronic hardware systemwide. There would be a Systems Analyst, and several Programmers. Operation and guarding of the equipment would be done by trained Computer Operators.

Electronic and mechanical maintenance would be performed under the supervision of a Foreman who administratively might belong to Maintenance (Power, Mechanical and Electronics), but would receive training and instructions from the Systems Division.

CONCLUSION

Systemwide coordination of operating procedures, train control facts, and computer programming, is a necessity for successful start-up and operation.

This function can, realistically, only be discharged by a separate group, headed by a person ranking equally with the division heads, and who possesses adequate qualifications.

(ATTACHMENT)

YARD INTERFACE PROBLEM

At Central we must know the exact procedures by which trains are received and dispatched at the yards, to ensure that the computer programs are compatible.

The yard interface has been a problem since I, in 1968, was commissioned by Westinghouse, as part of my training, to program Train Dispatching. It was obvious that neither BART or PBTB had clear concepts of the requirements of yard procedures for dispatch and reception of trains, and were unaware of the technical possibilities and cost of implementing good procedures.

A good deal of discussions with Transportation people, with PBTB and Westinghouse, ensued at different levels, and several Change Notices

> were issued, but the matter remains as confused as ever (reference my letter to Mr. Wargin written in frustration 4/2-69).
>
> When the latest Change Notice (D 120) on the subject appeared, I felt that a total analysis of the situation was mandatory, not only to solve the programming problem, but also to devise sensible procedures for the yard to follow, and suggest technical aids to facilitate operation.
>
> A report is now completed. It describes the provisions established in the Train Control and Yard Control Contracts, and the subsequent changes requested by Change Notices.
>
> Next, it points out discrepancies, errors, and omissions in the Notices.
>
> Then follows an analysis of yard operations (as related to testing, dispatch, and reception of trains) with suggestions of efficient means of communicating orders from towerman to the yard attendants.
>
> Finally, a rough cost estimate of implementing the system, is given. (There will be considerable savings, compared with the present provisions.)
>
> The report should be discussed, first with yard and transportation personnel, and, upon their concurrence, with PBTB, Westinghouse, and PHILCO. Finally, a sweeping Change Notice to settle all questions, should be issued.

The significant thing about this memo from Hjortsvang to Ray is that, for the first time, Hjortsvang goes beyond simply expressing his concerns about technical problems and the inadequate monitoring of and control over contractors. Instead, he proposes a solution: the creation of a Systems and Programming Division. This proposal adds a new dimension to the conflict between the engineers and BART management, by providing a new way for management to interpret complaints of disgruntled employees.

Management originally seemed inclined to view such individuals as employees "who didn't see the big picture", or as having personality traits which produced confrontations rather than compromise, or as being out-and-out troublemakers. But with Hjortsvang's proposal to create a new Division in BART, it became possible for management to view critics as self-interested and power hungry.

Sometime early in 1971, after Hjortsvang's futile attempts to get a hearing from management about his concerns, several events led him to attempt to go "outside normal channels" to present his problems to a Director of BART. The events were not planned, and Hjortsvang, although he knew of Directors who might be interested in seeing him, had not really been considering trying to get his concerns before the Board.

> I had no idea at that time (early 1971), or thought about meeting with certain members of the Board of Directors to voice my concern. I thought that it was important enough that the leadership of BART should know the dangers that were ahead... However, thinking about it, I realized that it might be a dangerous thing for me to do as a person because if I couldn't rely on these people's discretion, it might have bad consequences for me.

Hjortsvang described how the opportunity to contact Board members became known to him, and how this led him to try to reach Board members he felt would be interested in hearing from him.

> There were two people that obviously were suspicious of a lot of the staff's actions and critical about them, and one of them was Mr. Blake, a representative from San Francisco, and another was Mr. Bianco from Richmond... These were the people I had in mind. I was thinking about how to get in touch with one or the other of those and simply ask them what should one do when one had concern like this.

When asked if he discussed his thoughts or ideas about contacting Directors with anyone else, Hjortsvang replied:

> At that time, no. I simply--I had an opportunity--there was a man, a union organizer in there by the name of Gil Ortiz... He was at the time employed (at BART) as an electrician, but obviously he was there to organize labor, and he didn't hide that. And he had a lot of connections within the labor movement and, among other things, he said that these people were supported by labor and would be responsive to what--to approaches like I had in mind, and he said if I wanted, he could arrange a meeting.

Hjortsvang was asked if he approached Mr. Ortiz about a meeting with the Directors.

> I didn't approach him. He came around in the office and he went all over and talked with everybody and also with me about all kinds of things, and he--I didn't know how we got on the subject of that something wasn't right. I think it was common knowledge with even the workers, the technicians working in that place, that everything wasn't as it should be and something ought to be done. That was sort of in the atmosphere.

Hjortsvang went on to describe the concerns he expressed to Ortiz and which he assumed were the basis for Ortiz's offer to set a meeting with "labor-supported people on the Board."

> My concern was more with the management of BART than it was the technical problems. My reasoning was that if management is effective, they will recognize when there are technical problems and take the proper steps to correct them.
>
> So to me, all the time I was at BART, my thinking was that we had to wake up management to take some action, be aware of these dangers that I had.

> I did not try to, to a large extent, to point out specific faults in the design or dangerous conditions and so on because I felt that nothing would come out of that before we had the means to deal with such things.

Hjortsvang accepted Ortiz's offer to put him in contact with some people on the Board. (To Hjortsvang's knowledge, Ortiz had not approached Bruder about similar matters, and at this time, Blankenzee was not yet an employee of BART.) An arrangement was to be made for Hjortsvang to meet with Mr. Blake, one of the BART Directors. But the meeting never came off, because Hjortsvang changed his mind. Why? "Because after thinking about that, I, for a while, decided that I did not want to--to start anything, anything that could be dangerous for me personally."

In spring of 1971, several months after his discussions with Ortiz, Hjortsvang tried another approach. He set up a luncheon meeting with Robert Bruder and with Jay Burns, another BART engineer. Burns was an electrical engineer who worked in the Engineering Department with Bruder on the contracts. The reason for setting up this meeting has been described by Hjortsvang.

> Well, my thinking was that the department that Mr. Burns was a part of was the Engineering Department and it seemed reasonable to me that if anything had to be done engineering-wise about the problems that I saw that BART had, that might be the right organization to take care of it. It was at that time organized to supervise all civil construction and mechanical construction and Mr. Burns was an electrical engineer, so at least there were--there was the background, the rudiments of an organization that also could take care of the train control problems on a more efficient basis. And therefore, what I had in mind was whether they felt that I could approach--was that Mr. Tillman that was the head of that department?--to become transferred to that with the understanding that I would be concerned with train control

> problems and supervision of the installation of the train control system. It seemed to me to be a good way--The Maintenance Section that I worked in obviously was not interested in systems problems. They were essentially interested in preparing for maintenance of the system once it was delivered and I would like to be part of the Engineering Department where we could look at the functions and the design approach.

Bruder's recollection of the luncheon meeting puts greater emphasis on Hjortsvang's interest in using them as a means of access to their supervisor, rather than in seeking a transfer to the Engineering Department. As Bruder described it:

> Well, Holger came to us because he couldn't get anywhere in his organization to get things organized and on a path that would lead to a reasonable schedule and solving all the problems...and he wanted support because he looked at us as more professional, well Burns in particular had a professional ticket. I had a degree. We were more aligned to him as far as an engineering level and engineering work and experience. And he wanted support from us to go to his people or up to Tillman who would be the common point above Fendel. Somehow get information up to upper management that things weren't being taken care of.

In their court depositions of 1973, both Hjortsvang and Bruder stated that the luncheon meeting was not concerned with obtaining Bruder's and Burns' support of an approach by Hjortsvang to one of the BART Directors. However, in a 1977 interview, Bruder had a different recollection:

> I think at the time (the luncheon) even he (Hjortsvang) was talking then if we should go to somebody in the Board of Directors because nobody inside is listening. They don't want to talk...he wanted to go to some member of the Board. He wanted us to support him and go with him. To ring the bell or whatever you want to

call it. (Ed.: Bruder apparently means "To blow the whistle".)

According to Bruder, Jay Burns stated at this luncheon meeting that he didn't want to be involved. Bruder described his own response to Hjortsvang: "I said, I got a family with four kids. I've lost one job this way (Ed.: See Chapter 3). I don't want any part of it.'"

Hjortsvang has told how he felt after this failure to get support from Bruder and Burns. "I sort of gave up... I had to do what I'm supposed to do here and not concern myself with BART's management."

It appears, then, that at this point in the history of the BART incident, Hjortsvang and Bruder were totally stalled. They had been worn down by the contrived resistance of their superiors to being informed about and to responding effectively to the concerns of the engineers. In reaction to this invisible, but impenetrable wall thrown up by management, the two engineers seem to have decided that they had exhausted all possible means for reaching management about the problems they perceived.

The Arrival of the Catalyst

But in May, 1971, things began to change. It was during this month that Max Blankenzee joined the BART organization, became the catalyst who reawakened the, by then, latent concerns of Hjortsvang and Bruder, and set the three engineers on a course of action from which there was no turning back.

Since Blankenzee had already worked on the automatic train control system while he was a Westinghouse employee, the technical problems he encountered at BART were not new to him. His analyses of some of BART's technical problems drove him to the same conclusions as those reached by Hjortsvang, with whom he worked on a daily basis. Their common concern about the technical issues reached a critical stage after both attended a computer seminar given by the Advanced Institute of Technology on September 8-10, 1971. The two were assigned by management to attend this seminar as part of their duties, in the expectation that it would assist them in their work on the central control system.

On September 22, 1971 Max Blankenzee sent the following memorandum to his supervisor E. F. Wargin.

> The following evaluation of the software system has been based on the text given in the seminar by the Advanced Institute of Technology in standards of program development.
>
> The Central Control software development by Westinghouse has shown us enumerous problems when we tested the simulator. It became obvious when I evaluated the software system that the lack of enforcement in design procedures and standardization in this system is the major cause of these problems.
>
> The software system that we have in the present condition is not usable because of the following inadequacies:
>
> 1. The software system is not reliable because of high failure rate of the software.
>
> 2. It cannot be recovered quickly from a software or hardware problem.
>
> 3. The documentation on the system is very poor.
>
> 4. This makes the system impossible to maintain.
>
> 5. The software does not, in some cases, perform according to the real train operation.
>
> We can continue with listing all the software inadequacies but at this time I would like to propose to the management of BART a presentation showing what is wrong with the Central computer software, how it got into this condition, what we can do about the software system

> at this time and the troubles BART will encounter if we don't correct the present software status. I feel that this presentation would take a half-day of management time.
>
> I would appreciate your support in every way possible to correct the software status and to let BART management understand what trouble BART will encounter with this system if the software is not corrected. I also appreciate your comment in the near future concerning this matter.

In this memorandum, Blankenzee straightforwardly requests an opportunity to present his views on the problems of the central control software directly to BART management. He apparently received no reply to his request.

On October 1, 1971, Blankenzee sent a monthly progress report to Wargin in which he again notes the need to bring to the attention of management some of the problems of the central control system.

> <u>MAINTENANCE</u>
>
> The maintenance procedures and spare part list have been submitted and 95% is completed.
>
> Training lectures for the computer technician are being prepared and is 40% completed.
>
> <u>EDUCATION</u>
>
> September 8, 9, 10 - Mr. Hjortsvang and I attended a seminar given by the Advance Institute of Technology in standards in program development.
>
> <u>PROPOSAL</u>
>
> Transportation department submitted and proposed a program which could validate train schedules created by the scheduling department.
>
> Software test procedures submitted by PBTB have been reviewed and commented.

> PROBLEM AREA
>
> The Central Computer hardware and software documentation need to get management attention. Without correct documentation, it would be impossible to do any maintenance on both software and hardware.
>
> PLANS FOR NEXT MONTH
>
> Completion of training lectures, starting of training classes, simulator testing, finish of Maintenance procedures and spare parts list.

On November 5, 1971, Blankenzee sent two more memos to Wargin. The first one states that the scheduled tests of several of the BART systems to be undertaken by Westinghouse and PBTB should be reviewed by the computer group at BART (i.e. Hjortsvang and Blankenzee), and that, moreover, there should be no acceptance of these systems as complete without approval by a BART computer engineer.

Blankenzee's request was designed to give BART, rather than PBTB, the final say as to whether documentation and test of a system was adequate. This is the same position taken by both Hjortsvang and Bruder many times before Blankenzee joined BART.

The second memo of November 5, 1971 pointed out the inadequacies of the Westinghouse Training Course, designed to enable BART computer technicians to maintain the computer system once it was turned over from Westinghouse to BART.

> With the pre-revenue testing coming up in the next few months, several systems in BARTD will undergo witness and acceptance testing.
>
> It should be realized that none of the witness testing and computer system testing should be done without having the documentation and test procedures completely reviewed by the computer group.
>
> It should also be clearly understood that to prevent acceptance of non-computer systems,

> none of the testing should be done without having a computer engineer participating as a BARTD witness.
>
> We have noted that the training for the computer systems is being accepted without reviewing of the course contents. To make sure that we get the desired quality of training, all training text should be reviewed by the computer engineer before we accept or attend any training courses from the contractor.
>
> With the continuation in the scheduled training course for the computer technician, the following should be taken into consideration:
>
> Westinghouse Training Course is only an introductory training to the P-250 computer system, and anyone who has completed this course is not qualified to do any maintenance on this computer system.
>
> Attached to this letter will be an outline of the minimum training a technician should receive before he is qualified to perform maintenance on a computer system.
>
> As you can see, additional training should be provided for the technician before he does any maintenance on this system.
>
> If this additional training is not provided, the computer group and technician cannot be held responsible for any failures that might occur because of improper maintenance performed on the computer.

Just as many of Blankenzee's concerns about technical problems did not differ substantially from those expressed earlier by Hjortsvang and Bruder, the reaction from his supervisors was also a familiar one. Blankenzee recalls the response of Wargin to his memoranda.

> Mr. Wargin never responded, only verbally....I cannot state them exactly as how he exactly worded them, But more on the trend of: Well, we know we are getting all these problems, we

> know we have all these problems, don't worry about it. That shouldn't be any of your concern... And also, that I shouldn't criticize Westinghouse as much as I did, that it was for my own good. If I just do the things, wait until the whole thing was accepted and then try to correct these things.

Blankenzee stated that he didn't agree with Wargin's view that we should take things as they are and try to correct them later. He has explained why he disagreed.

> Because I felt we were trying to get a product as being the customer. BART is the customer on the Westinghouse contract. We should try to get not only our money's worth, but the best kind of equipment that we could get for the total system.

Blankenzee's assessment of BART's problems included two new elements not raised earlier by his colleagues. First was Blankenzee's view of the structure of power within the BART organization. He appears to have developed a much more politically sophisticated analysis of why he and his colleagues were continually unable to reach top management with their concerns.

> Ed (Wargin) was a nice soft spoken man, but I think that he had the God-given fear in him, put in him by Charlie. Charlie Kramer who was Ed's boss was a very forceful, very agressive individual, a very strong individual. He didn't want to hear about a lot of stuff...
>
> And there was political conflict (between the maintenance and engineering staff)...and Ed didn't want to seem too much of a troublemaker, I think. I remember an incident where Holger wanted to go talk to Ed, and then Holger wanted to talk to Charlie, and he requested to talk to John Ray or higher up, and Kramer said, "No way". You don't need to talk to him. I don't think its necessary. You know, I'll decide when its necessary..."
>
> Kramer is a very powerful strong man that does not like to get the information. He wants to

> keep the information in his shop and that's it. Because nothing that had to be related to John Ray and to Stokes, and those people, came over out of Kramer's shop without Kramer approval.

There is no doubt in Blankenzee's mind that the middle levels of management, from Wargin to Kramer to Ray, were aware of the problems that the engineers had been trying to bring to management's attention. He is less certain about whether the general manager, Mr. Stokes, had knowledge of these problems. On the one hand, Blankenzee is skeptical of Stokes' statement that the first time he heard about the problems with the automatic train control was when he saw an anonymously circulated memorandum of November 18, 1971 concerning the technical and management problems at BART (the memorandum was written and circulated by Hjortsvang). On the other, he suggests that it is possible that Stokes had not heard about any of the problems.

> You know, there was stacks of reports going, (pause) unless Kramer kept them away from Stokes... I don't know what Kramer's report was to John Ray and what John Ray's reports were to Stokes, but the reports sure came into Kramer's shop and sure as heck the guy from PBTB.

It was this ambiguous feeling about which side Stokes was on that later led Blankenzee to join in one last effort to reach top management (i.e. Stokes) through the Directors rather than through the management chain of command.

The second new element in Blankenzee's assessment of BART's problems was the view that the people who would ultimately bear the burden of the contractor's failures to deliver finished product (i.e. a completely tested and documented train control system) would be the BART engineers and technicians who would have to maintain the system. In other words, he believed that BART management's view that the engineers should just stand by until Westinghouse delivered their product and then show their ex-

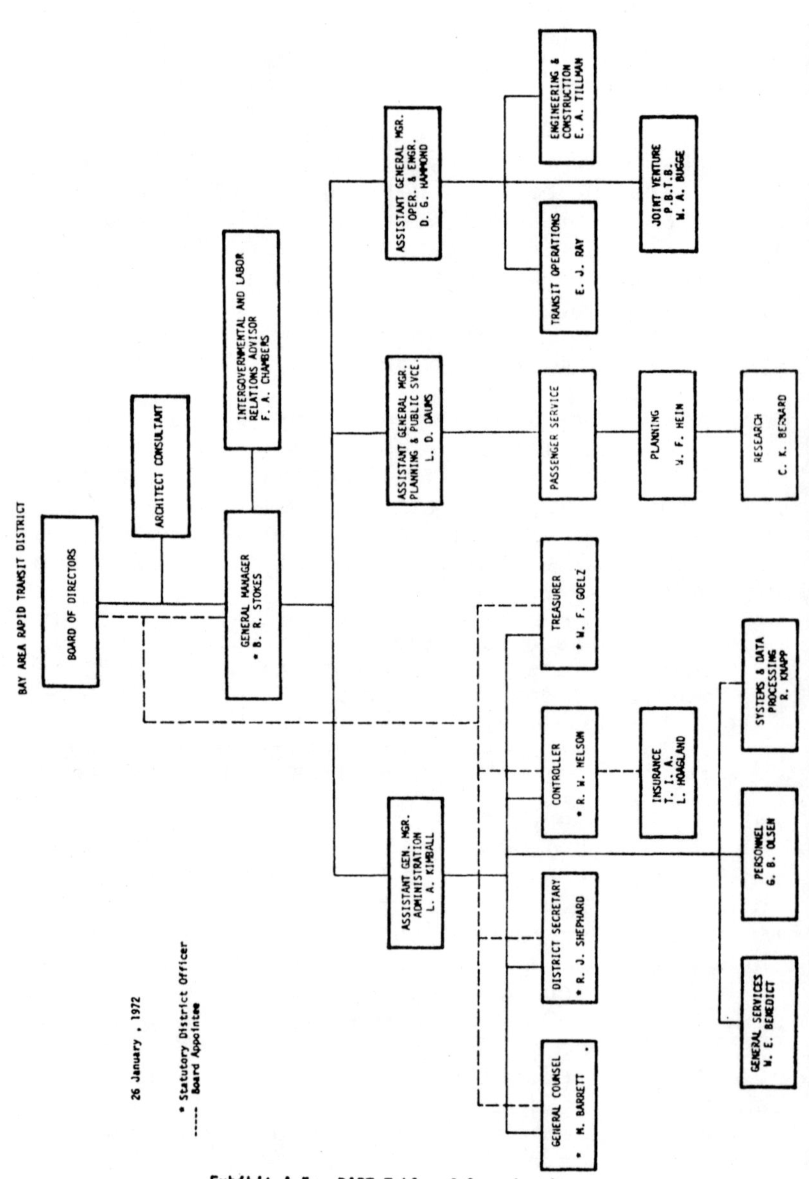

Exhibit A-5: BART Table of Organization

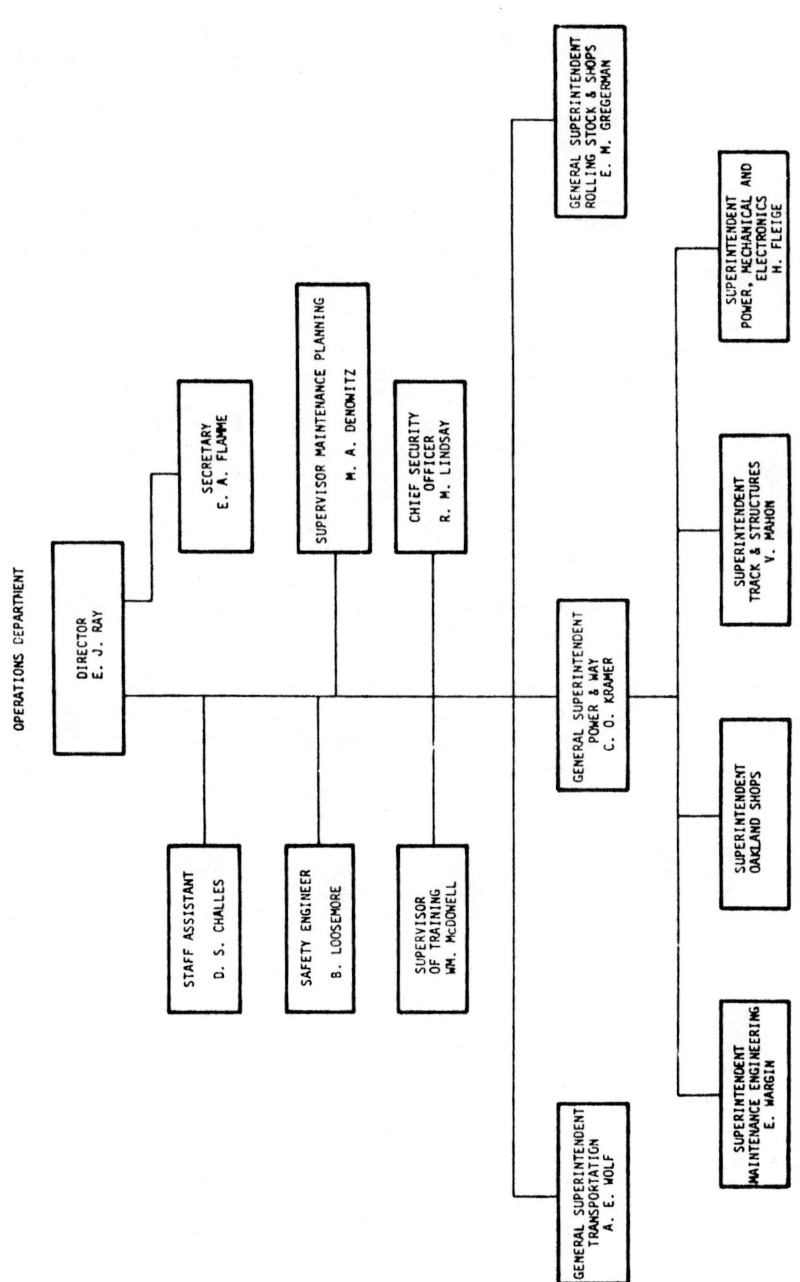

Exhibit A-5 Continued

pertise and clean up the problems, was not only unacceptable as a management policy, but also guaranteed tremendous problems for the working engineer and technician.

Blankenzee describes these concerns:

> You see, the biggest fear that we (had), that came creeping up, that if we could operate the system and end up being responsible for the system, what are we going to maintain it with? We got no knowledge, there is no documentation and how are we gonna find the problems. There's gonna be an extensive amount of work, plus how are we sure that what we are doing is right, because we didn't know the design, we don't know the internal workings of the system. At that time, Westinghouse went out knowing that this would happen...they were supposed to train the technicians and engineers and at the turnover time, we were supposed to carry over our own system.

Blankenzee gives a specific example of the problems experienced by BART personnel.

> We had a simulation system that's supposed to show all the working of the simulator, how the train would operate from one end to another going on the route, and the operations people of BART were going to get trained on this thing. When Westinghouse was done with the coding and they had it in the computer system, the software division walked out of there. And Ken Dale said, "Hey, you can't go yet." And they said, "Well, what do you mean. We're done." And they left. They flat went home, and Ken had to fight like hell, political-wise, to put the pressure on. Because when they had to come up with a demonstration of this thing, nothing worked.

Finally Blankenzee summarizes his comments about how the BART people would have to "pay the price" for BART management's failure to insist on a finished product from the contractors:

> There were two thoughts going into BART. One thought was, let the contractor do it. They

> are responsible. If they foul up, afterwards we will sue them. Another thought was, and this was mostly Kramer, forget about the contractor except know whatever is wrong we'll fix... We thought that neither of them were right because if you don't have the right documentation, the right training, if you don't know what went into the system, how can you fix it?...and the people that would have been left responsible for it were the people like Wargin to fix it. And this is what we were trying to bring to Wargin's attention. That it will be your neck. But Ed knew that he would retire before the train system ever opened up so he wasn't interested.

But many of the other engineers and technicians at BART would be there when the system opened up. Therefore, Blankenzee's concern that he and others would inherit a system they didn't understand and could not maintain struck a responsive chord among BART personnel.

The result of Blankenzee's assessment of the BART situation was a reawakening of concern in Hjortsvang and Bruder, and a renewed interest in taking direct action to deal with the problem. In addition, and perhaps more significant, was what the three engineers perceived as an increased sense of concern among a broader base of BART engineers and technicians, giving them the feeling that there was growing support among their colleagues for their position and the actions they would ultimately take.

The period between September and December of 1971 was one of widespread and intense discussion among BART engineers about technical and management problems. The initial stimulus came from Blankenzee, who repeated the earlier efforts of Hjortsvang and Bruder to reach management by written and oral communications. There were frequent discussion of BART's problems among the three engineers, and between Blankenzee and other engineers and technicians at BART.

When Hjortsvang and Bruder were asked why they had changed their minds about trying to do something about BART's problems after having given up on the whole affair earlier, they answered as follows (first Hjortsvang, then Bruder).

> Well, what happened later was that Mr. Blankenzee was hired and I talked with him and the problems and I suppose that that was what made me interested again in trying to alert BART as to what I felt was wrong.

> Well, I guess strength in numbers or something, I don't know. I really have no reason for the change outside of that, plus more time it got on and less had been done, things were getting worse, more things were being said to the public that still didn't stand the test of facts...

Hjortsvang's renewed interest led him to distribute an unsigned memorandum on November 18, 1971 to as many levels of the BART organization he could reach. The memo, simply placed on the desks of BART employees prior to their arrival at work, reintroduced the proposal for a Systems Engineering Department that Hjortsvang had made in his signed memo of January 7, 1971 which he had tried to get to John Ray. The memo of November 18, 1971 reads as follows.

> Train operation and control is a technical matter which definitely requires leadership by competent engineers and programmers.

> In the BART organization chart you will find no department to coordinate designs, determine performance standards, correlate with contractors and PBTB in technical matters, and in general assure the best possible system.

> As a result, BART has not had proper influence on the design and construction of the automatic train control. There has been no way of evaluating the Contractor's methods and progress. Unfortunately, PBTB's representatives have not done much better. For example, only after Westinghouse's central control development showed alarming signs of being deficient

and late, did PBTB assign a competent computer control engineer to look into the work.

The lack of technical competence and an undue reliance on PBTB's judgement, is the direct cause of many costly mistakes and delays, even accidents.

With the growing pressure to get the system started, the temptation to accept equipment in spite of deficiencies and lack of proper documentation, in hope of later correction, makes technical supervision more important than ever.

It is difficult to understand that BART's management has not realized the necessity of a technical department for train control and operation, since the organization at industries using computer controlled processes invariably have a Systems Engineering Department in one form or other.

This is confirmed in Bulletin 1658, 1970, by the US Department of Labor, which reports on Computer Process Control and Manpower Implication in Process Industries. None of the installations surveyed is as complex as the BART system, but the organizational implications of computer control are well illustrated. Anybody concerned about system operation should read this publication.

In order to create an entity within BART to deal with our technical difficulties in a professional and efficient way, it is proposed to establish a Systems Engineering Department immediately. Most of the personnel can be extracted from existing departments, but a number of new specialists should be hired. In this process there will be an opportunity to examine the other "engineering departments" to determine their adequacy.

At just about this time (November-December, 1971) Blankenzee held discussions with nine other resident engineers and technicians at BART about his concerns. He felt that the nine were very sympathetic to his view because they faced the same problems he did, and he describes what he felt transpired during these discussions:

> What I did was I said, "If we are all concerned we should stand up as a team and fight for this." And they said, "Yes, we all agree." And within our conversations it has been mentioned many times, "Yes, we are very concerned. We wish somebody would do something about it."

And somebody did.

The Approach to the Directors: Circumventing Management

The idea of trying to reach members of the BART Board of Directors came up in much the same way it had when Hjortsvang considered the idea at least six months earlier. Gil Ortiz, who had periodic contact with Hjortsvang and Blankenzee during lunch hours, after work or by telephone, had once again extended an offer to put them in contact with "higher-ups" in BART. Blankenzee recalls Ortiz's comments:

> And Gil said, "Well, look,...if you want me to, I can set things up for you to talk to him, and they'll make sure your name won't come out," and we said, "Well, we don't want to get canned for this." He says "Let me tell you something. These people are very good, and they are very much concerned about you and about BART, and they have the same concern for the public."

Ortiz specifically mentioned the names of Directors Bianco and Blake. Several weeks went by with Ortiz doing little but making periodic calls to indicate he wasn't able to make contact and would keep trying.

Sometime in early November of 1971, though, Blankenzee did succeed in having a telephone conversation with Mr. Blake. He reports that: "I told him that we engineers of BART had a concern and that we would like to see, like to present this concern to him and would see if maybe he could talk to the upper management and be able to make them aware of the problems that we had."

Blankenzee reports that Blake responded, "Fine, go ahead. You set up a meeting and I'll be there."

The meeting was never set up. Blankenzee recalls that he was unable to reach Blake again by telephone, and that Ortiz was unable to contact Bianco. While Hjortsvang was aware of the steps that Blankenzee and Ortiz were taking to set up a meeting, Bruder apparently was not. His only knowledge of what was going on was a phone call from Hjortsvang who said that "Max had a contact with Mr. Blake," and asked Bruder if he would come to a meeting. Bruder said yes.

It is not clear whether the intended meeting with Blake and Bianco did not take place because of inability to make the contacts, as Blankenzee recalls, or because of other questions and problems that led to a deliberate decision to delay that meeting. Hjortsvang and Bruder both state that they were worried about trying to get their ideas across to non-technical people. It was this concern that led to the decision to bring in an outside consultant who might strengthen the case the engineers hoped to put before the Directors.

Hjortsvang describes his views on why the meeting with Blake was off.

> I had some concern. I felt that if you try to explain the difficulty, technical difficulties, to unprofessional people, you have to have somebody that can sort of back you up and tell that what you say makes sense, and it's common practice to do such and such a thing and, for that reason, he (Blake) discussed with Blankenzee whether we would find a man that was familiar with computer control systems and could therefore understand our concerns and understand the technology--the terminology you would use when you explain these things to other people, and I felt that having a person at the meeting that just would serve that purpose would create more credibility and confidence in what we said.

Bruder's statement on the matter makes it clear that it is Blake who felt that an outside consultant should be obtained before any meeting should take place.

> Because we saw the same basic problem we can't communicate with some of our superiors that don't have the specialized experience or knowledge how are we going to communicate to even a relative layman that isn't even technically or engineering knowledgeable. And there were attempts to have him (Blake) have a consultant, you know, to be an interpreter for him.

The Burfine Report

For these reasons, Blankenzee contacted an independent engineering consultant, Edward Burfine, to confirm the engineers' views when they were presented to the Board. Blankenzee describes how he thinks the decision was made.

> (The idea to get a private engineering firm) is what came back from the conversation, I imagine, between him (Ortiz) and Blake and Bianco. In my phone conversation with Mr. Blake I stated, I said, "We do not have any--we do not have any backup whatsoever, but if you do want to, you can get some consultant to confirm what we say, and could we, possibly, get a consultant." And he says, "Go ahead. Arrange everything and I'll be at the meeting."

Blankenzee goes on to describe his initial conversation with Burfine, who was recommended to him by someone in the industry.

> I contacted Mr. Burfine and told him that on behalf of the Directors that he was supposed to be there to pass judgement on our statement of BART. Bias or non-bias feeling from what we had said, from what condition BART is in, so that he could confirm if we were right or... (sentence is not finished).

A short time after this call, Burfine came to the Lake Merritt office of Blankenzee and

Hjortsvang one morning in December, 1971. Bruder joined them several hours later--around 2:00 or 3:00 in the afternoon. Burfine listened to the views of the three engineers about BART's problems.

Hjortsvang later reports some concern about what transpired during these conversations with Burfine:

> he knew computer control systems and that we would like him, as a specialist, to talk about it, but not let him go into any specifics about BART. Nevertheless, he started asking questions about the system, to my consternation, because I didn't want to discuss BART's internal things with him at all. And I told him several times when he questioned these things that I didn't think they were of any interest in connection with this meeting we were going to have. But I--I could not stop him mostly because I think Blankenzee felt that he should at least accommodate him and answer his questions and tell him about what we felt was wrong.

As a result of his visit to the Lake Merritt facility and his conversation with the three engineers, Burfine produced a seven page report entitled "Review of BART Operations." It was dated January 12, 1972 and it was stated on the report that it had been requested by two members of the Board of Directors.

Burfine sent the report to Blankenzee and Hjortsvang, neither of whom expected to receive such a document. In fact, Hjortsvang states that he specifically requested that Burfine not write a report. Specifically, according to Hjortsvang, Burfine, upon completing his one-day visit to the Lake Merritt facility had stated: "I'll go home and write down a summary of what we talked about." To which Hjortsvang had replied: "No, Mr. Burfine, don't do that. That's not necessary."

In addition to being responsible for Hjortsvang's professional distress, Burfine's report caused a different kind of consternation for Blankenzee. Burfine sent Blankenzee a bill for $500. Blankenzee had understood from Gil Ortiz that the bill would be paid by Bianco or Blake or some other Director. And so, on the basis of this understanding, he refused to pay the bill. Thereupon Burfine, according to Hjortsvang, threatened to send it directly to BART management. (The disposition of the bill is still unclear. Blankenzee claims that the bill was finally handled by Director Daniel Helix after a meeting during which the engineers explained the problem. Helix, however, claims that he did not pay the bill.)

In any event, after all this effort to get an independent consultant to back up the engineers when they would meet with Bianco and Blake, that meeting never materialized. None of the three engineers seems to know exactly why. But when it became clear that there would be no meeting with the two Directors, Blankenzee was again in touch with Ortiz. According to Blankenzee, Ortiz responded by saying. "I have a Mr. Helix who is interested in listening to you. I'll set up a meeting if you want to talk to him."

And he did. For the first time, the engineers met with a member of the Board. In early January, 1972, (late December, according to Helix) Hjortsvang and Blankenzee met with Helix in a local union headquarters in Oakland. Hjortsvang has described what transpired at that meeting:

> My first comment was that this meeting was confidential and I pointed out to Mr. Helix that we had some concerns about BART's management, and we would like him to know about it and use--perhaps use his influence as a director to get these problems solved in the interests of BART and the taxpayers.
>
> But I pointed out very strongly that I had a family and I was sixty years old and I was not going to expose myself to any open fight against BART. And he assured me that he would respect that and that nothing would happen to me or Blankenzee.

Blankenzee's description of the meeting is as follows.

> We told Mr. Helix that we had some concern and we felt that Mr. Wargin and our direct managers--Mr. Kramer and his manager-- were not interested in our problems at all and that we would like to get Mr. Stokes and the top management and let them know of our problems, hoping that somewhere we could make them aware. There are problems within the central control system and within engineering... And then we went on to ask Mr. Helix just to keep these things to himself and try to get in contact with the other members or the General Manager of BART.

According to the two engineers, Helix was very interested in what they told him and expressed his intention to help them. Hjortsvang gave Helix some copies of memoranda he had written about the need for a systems engineering division. In addition, Blankenzee told Helix of Burfine's visit to the BART facility and of his proposed consultant's report.

Helix told the engineers he would bring the matter before his fellow Board members and that he would also distribute copies of Hjortsvang's memoranda to them. Hjortsvang once again cautioned him about protecting the identity of the engineers.

> I said I hoped that he would not create a situation that would bring my identity in the limelight and he assured me at that time that even if this happened and BART went ahead and fired us, he would raise such a stink, he said, in all the three counties and all the 13 newspapers they would be sorry they ever did it.

Unfortunately, either Helix, a member of the Board for only three or four months, was led by his inexperience to making assurances which he could not realistically live up to, or Hjortsvang misunderstood what Helix told him.

The Politicization of the Controversy: The Engineers Get Lost in the Shuffle

Two days after the meeting between Hjortsvang, Blankenzee and Helix, the Contra Costa Times published a story about BART's internal problems in which Hjortsvang's memoranda were reproduced in their entirety. Hjortsvang was shocked and troubled by this event since he had had no indication from Helix of his intention to "go public" with the information. Hjortsvang's memoranda were also distributed to the Board members, and shortly thereafter the "Burfine Report" was distributed by Helix to the Board and to BART top management.

Hjortsvang promptly told Helix of his opposition to publishing these things in the Contra Costa Times.

> I said I felt that this was a sure way to defeat our purpose, to publish things like that. Because if there is something that ought to be corrected inside an organization, the most effective way is to do it within the organization and exhaust all possibilities there...you might have to go to the extreme of publishing these things, but you should never start that way.

The immediate consequence of this release of information for the three engineers was an attempt by BART management to find out who was leaking information to Helix and the Contra Costa Times.

Blankenzee was called in for a meeting with his supervisor, Mr. Wargin and was asked what he knew about the Burfine Report. He replied, "Nothing." He was also asked what he thought about the allegations in the Report. Blankenzee reports his reply:

"I said, 'I think that it was all true, and so did they.' Mr. Wargin said, 'We know all these things are true, Mr. Burfine isn't telling us anything new, but who contacted him?'"

According to Blankenzee, Wargin then asked him to "keep your ears open. If you hear from anybody that contacted--that could have been involved with Mr. Burfine, tell us about it."

The very next day Blankenzee was called in for a meeting with Mr. Ray, two echelons above Wargin in BART's managerial hierarchy. The purpose of the meeting, according to Blankenzee, was to identify the troublemakers within BART. "These were confidential meetings held inside that said if I had any knowledge or I could call any names, that I would not be labeled as the guy that pointed the finger at the other guy."

In response to Ray's questioning, Blankenzee again denied any knowledge of the Burfine Report. He was then asked by Ray about what could be done to correct the things alleged in the Report.

> We talked about a systems engineering group and we talked about a computer group and we talked about the--part of being able to work correct test procedures and be able to check things out and make recommendations that had a little bit of authority behind them and that would count within the changes in such a way that they were to the benefit of BART and not to the benefit of PBTB or Westinghouse.

Blankenzee says that Ray asked why, if these problems were true, people had never come to him to report the problems. Blankenzee responded that "if we came to him directly without their knowledge or tried to come to him, Mr. Kramer more or less told us that unless he approved we couldn't talk to Mr. Ray. And if we did talk to Mr. Ray, that we might as well find another job out on the street."

Hjortsvang was also called in for a meeting with Mr. Ray, who, he states, asked him: "Do you know who is involved in these things that we read about in the papers?" Hjortsvang replied: "You know very well I cannot answer your question," meaning that he would have to betray a confidence to answer it. Hjortsvang states that he did not deny his involvement with Helix and Burfine, but that he simply refused to answer whether or not he had any involvement.

At about the same time, Bruder had a meeting with Mr. Fendel, his superior, once removed, at which he also was asked if he knew Mr. Burfine, or knew of anybody else that might have talked to him. To the first question Bruder answered "No", and to the second: "It could be almost anybody working on the contracts in any level."

Immediately after his meeting with Fendel, Bruder contacted Phillip Ormsbee, BART public relations director and aide to Stokes, and made an appointment to meet him in his home in the evening. Bruder complained to Ormsbee about the efforts being made by management to find out who was working with Helix and Burfine. He said to Ormsbee:

> **Well, essentially I think BART has got worse problems in the technical management, and it's the personnel approach they are using in the situation. Well, just the Gestapo-type actions; calling people in and interrogating or trying to worry about who is releasing information rather than what does the information say, especially to the people involved professionally in it.**

Ormsbee listened to Bruder's concerns and asked him if he was involved in the Burfine affair. Bruder said he was not involved. Ormsbee then asked Bruder if he wanted to see Stokes, the general manager, and Bruder said yes. The meeting with Stokes was set up by Ormsbee and took place one or two days after the Bruder-Ormsbee meeting. Bruder describes what happened at his meeting with Stokes, which was also attended by Ormsbee.

> Well, we just generally discussed the lack of a controlled schedule to get the thing done on the target dates that we were telling the public and the interfaces between the rest of the system and the train control and the station wasn't being taken care of, and I couldn't get any response anywhere to get it taken care of.

Stokes' response is described by Bruder.

> Well, he brought up one schedule of how everything was plotted out, how we would get to the end point, which was more of a general chat of all the things everybody in BART had to do, not so much the technical contracts we were discussing, and he said he would set up a meeting. I guess he did. I don't know whether he told me then or he called me the next day and said he wanted me to meet with Mr. Hammond and discuss these things.

The very next morning, Bruder met with Hammond, Tillman, and Stokes. Bruder repeated his concerns about the problems he discussed in his meeting with Stokes and Ormsbee. Bruder summarizes the response of the top management as follows: "They said, well, we'd get back with Mr. Fendel and get this thing going."

During this period that BART management was engaged in its efforts to find out who was involved with Helix and Burfine, Helix was making plans to have Burfine attend a meeting of the Board of Directors and make a presentation about BART's problems. For this purpose Helix called a meeting in his Concord office which was attended by Bruder, Blankenzee, Hjortsvang and Burfine. Blankenzee describes the purpose of this meeting.

"The things that were discussed, how could we most effectively make the Board of Directors aware that there is a problem and what are the things that are definitely within BART that we could show to the Board of Directors that it is a problem with the train control system."

Working point by point through the Burfine Report, the engineers tried to prepare Burfine and Helix for the counter-arguments and questions which would inevitably be raised by BART management in support of its position (and in refutation of Burfine's assentions). The engineers also provided documents and professional writings about systems engineering to help Helix better understand their position.

The result of this meeting was a twelve page document, entitled "Importance of the Central Control System", that was to serve as the basis for the presentation before the Board by Helix and Burfine.

Of the three engineers involved in preparations for the presentation to the Board, only Hjortsvang states that he had reservations about, and sometimes opposition to, this decision. Hjortsvang felt, as pointed out earlier, that he could not fully support the decision to write and release the Burfine Report, largely because it did not support his position about the need for a systems engineering division. Why did it fail to support his position?

"By its mere existence. I think a report that is based on conversation with a couple of engineers lasting only a few hours could not support it. It could only be detrimental to my conclusions."

Hjortsvang, therefore, prepared a memorandum for Helix and Burfine entitled "Train Control--BART's Hangup" which he hoped would reverse the decision to have Burfine go before the Board, or at lease get him to change his presentation.

> I was afraid that if that man (Burfine) went up to the Board of Directors and presented his viewpoints, that he would fail miserably because of lack of knowledge of the system... (My memo to Helix and Burfine described) a more positive and more competent approach, and the--I hoped that reading it would convince Helix that an approach like this, and not discussing the details that Burfine proposed to discuss, would perhaps give rise to some understanding of the Board members that--how the situation was. Burfine was negative, and I tried to be positive...it (Burfine's approach) would be ineffective, because based on that superficial knowledge that Burfine had, it would only be damaging, and he would sure as hell be shot down when he started to discuss BART with the PBTB or with Westinghouse, or those people that had... (sentence is not finished).

Nonetheless, Hjortsvang did try to assist in the preparation of Burfine and Helix for the

Board meeting. He felt that, having stated his view, the matter was out of his control and that the ball was in Helix's court.

Between the time of this general strategy session and the planned appearance before the Board, Helix called another meeting. In attendance were Hjortsvang, Blankenzee, Burfine, Helix and two members of the Board, Bianco and Clark. Helix stated that it was his intention to have Burfine appear before a subcommittee of the Board--The Engineering Committee-- in advance of his meeting with the full Board of Directors. Helix showed the group some of the materials that had been prepared by Burfine, and some of the presentations he would make and the questions he would raise. It was, in essence, a "dry run" of what Burfine would do at the Engineering meeting.

Confronting the Board: The Big Guns and the Foot Soldiers

On February 22, 1972, Burfine made his presentation to the Engineering Committee of the Board. Although they were not in attendance, Blankenzee and Hjortsvang have summarized what happened at the meeting. First Blankenzee:

> Burfine was slaughtered. Westinghouse came around with a beautiful presentation in which Mr. Woodrow Johnson said that there was no problem with the computer system, that since he has been on in January--here it is-- in which Mr. Johnson made all his beautiful comments to the Board of Directors in which he stated there is no problems with anything, that Westinghouse had everything under control.
>
> And all the credibility was given to Mr. Johnson and to what Mr. Stokes had to say, because he made a presentation, too, and what PBTB had to say, because they made a presentation. And then this last one, Mr. Burfine tried to make a presentation, he didn't even get a chance to speak.

Hjortsvang's comments on Burfine's meeting with the Engineering Committee are much briefer, but no different in substance from Blankenzee's remarks.

"It turned out that he was just cut down to nothing at that meeting."

The day after the meeting of the Engineering Committee, Mr. Ray telephoned Hjortsvang and asked if he and Blankenzee would look at the Westinghouse Report and give their views on its contents. Hjortsvang sent his reactions to the report in a letter to Ray. Blankenzee has summarized the contents of Hjortsvang's letter:

"(Hjortsvang wrote that) Westinghouse was giving BART a good snow job, that if management believes everything that Westinghouse said in there, we really were in trouble."

Blankenzee chose not to comment himself on the Westinghouse Report, as Ray had requested, and he explains his reasons for not doing so.

> In my opinion, I thought if any highly intelligent men in BART knows what goes on in his own facilities or the qualifications that Mr. Ray and Mr. Kramer had, they knew that it was a whitewash. I mean it was so obvious that I figured I didn't have to comment on it any more.

After the failure of the Helix-Burfine presentation before the Engineering Committee, Helix had a telephone conference call with Hjortsvang, Blankenzee, and Bruder during which they concluded it was time for the engineers to come forward and present the facts to the Board. Since Burfine did not have all the facts and could not make an effective presentation to the Board, the only choice left to the engineers was to "go public" or give up the battle. They agreed to do so, and also to try to get their engineering colleagues to join them.

Blankenzee describes what happened when they tried to get support for their meeting with the Board on February 24, 1972.

> What happened was the day before or the day that we were supposed to have informed--to go before the Board of Directors, we called all the engineers together and we said, We are the engineers that have contacted the Board of Directors. We would like to have your support. Would you all go with us to the Board meeting?' And everybody laughed and said, "You got to be kidding." And so we were stuck.

Seventeen engineers were solicited by Blankenzee, Bruder and Hjortsvang for their support. Their reactions are paraphrased by the three engineers as follows: "Well, you guys are great heroes, but we want to feed our families."

Blankenzee, Bruder and Hjortsvang never did get a chance to tell their story to the Board of Directors. Helix called the engineers the afternoon or evening after the Board meeting and, according to Blankenzee, reported to them as follows on what had happened at the Board meeting that day:

> Well, I'm ready to bring my engineers forward--I think it was Mr. Dougherty, I'm not sure. It must have been in the records as well--no. Mr. Anderson came forward. He said, "Well, let Mr. Stokes take care of the engineers and let us just decide this matter in a state of confidence, and if we have a confidence vote from Mr. Stokes and Westinghouse." We said, "Aye", and there was a vote made on that, and it was 7 to 5 vote or 7 to 1, whatever it was, and it was done with. And in the meantime we had told everybody who the engineers were...

Management Retaliates: The Engineers are Fired

On the morning of March 2, 1972, Holger Hjortsvang received a call in his office from Mr. Ray's secretary. Mr. Ray wanted to meet with him. When he arrived, Hjortsvang found himself with Ray and Charles Kramer. Hjortsvang describes what happened.

> I think Mr. Ray spoke first and said that he was unhappy to tell me that it had been decided that my service was no longer required, and handed me a paper, or rather two papers, indicating that I would have the choice between voluntarily--what's the word-- resigning, or being dismissed. My response was "Why?" He repeated that my service was no longer required, and that it was not in his hands. It was a decision from the General Manager. I said, "What is it I am accused of, since you fire me, or since you dismiss me?"

Ray simply repeated what he said earlier, giving no reasons for the action. Hjortsvang refused to resign, but Ray encouraged him to think about it again because there might be advantages to resigning rather than being fired in terms of retirement benefits. Hjortsvang reasserted his intention not to resign. Ray then asked him to sign a receipt for his salary up until that day, and for vacation pay. A security officer then accompanied Hjortsvang to his office so that he could gather his belongings and leave the premises.

On that same morning--March 2, 1972--Blankenzee was meeting with a BART contractor in the Alameda Yard. He called Hjortsvang at 11:00 a.m. to tell him when he would be returning to the office. Hjortsvang said that he had just been fired and that they were looking for him (Blankenzee). Blankenzee returned to his office immediately and later reports on what he observed.

> The head of security at that time was also in Mr. Hjortsvang's office and so was Mr. Kramer's assistant in his office, too. And he was trying to pack his boxes. I guess the head of security left for a couple of minutes. I don't know what he did. He went to call Mr. Ray because I got a phone call after that and Mr. Ray says--his secretary says, "Mr. Ray would like to see you in his office." He brought me to Mr. Ray's office and the head of security...waited outside the door. Mr. Ray was in there and Mr. Kramer was there, and Mr. Ray says, "I have very bad news," he says, "I want you to know",

> he says, "that we have to ask for your resignation or we have to terminate you." I says, "What's this all about?" And he said, "Well, this comes as an order of the general manager." I says, "Then I want to see the general manager." He says, "The general manager is not available." I says, "Well, that was very convenient, wasn't it? But what's it related to?" He said, "Well, to your affiliation with Burfine, I guess." I said, "Well, do you know?" He said, "No. This has not been my order, I want you to know that," he says, "this came from the general manager." So I made a statement: "Well, we are still living in a democratic society, and we will meet in court about this matter...You can't fire me like that without giving me any response and not talk to the people who were doing the firing."
>
> I was subsequently guided down to my office and it was something like 11:30. And he (security man) said "You should pack your boxes..." And I said, "Well, you know, it's 11:30 and I want to go to lunch." He says, "No. You are going to pack here and then you are going to leave the building." I says--well the things that I said weren't too good in the conversation anyway. So I went to lunch. I had to turn in all my papers and I could not get into the building unless the security guard met me again in the front of the building and they let me through the office and a security guard waited in front of my office. The head of security was in my office, and so was Mr. Kramer's assistant there, and when I started packing my boxes, they said, "Just a minute." And they went through all my office and said, "This you could take and this you couldn't take." So I packed the things that they allowed me to take, and then I left.

The next morning, March 3, 1972, Robert Bruder was called in for a meeting with Mr. Fendel and Mr. Tillman. The meeting started with Tillman asking Bruder a series of questions. Did you have any dealings with Mr. Helix or Mr. Burfine? Did you ever discuss with Mr. Helix the question of appearing before the Board? Did you ever try to recruit other

BART employees to support your position? Bruder's answer to all three questions was, "No."

According to Bruder, Tillman responded by saying he had reason to believe that Bruder was involved in these matters and that he was, therefore, obliged to ask for his resignation. Bruder refused to resign and, thereupon, Tillman indicated that "we will have to let you go." According to Bruder, Tillman apologized for not having his check ready. Bruder left, with Fendel driving him back to his office to pick up his belongings.

The firing of the engineers was the culmination of a series of events which had started well over a year earlier. The series began with the independent recognition by Hjortsvang and Bruder that there were management, contracting and technical problems at BART which were not being adequately addressed. After a number of futile attempts to go through channels or to reach the Directors, the two engineers were essentially reconciled to living with the problems and forgetting about trying to do anything about them.

The next stage was inaugurated by the appearance of Max Blankenzee in May, 1971. Blankenzee saw the same problems and had the same concerns as Hjortsvang and Bruder had earlier, and he stimulated them to join with him in another attempt to set things right. This attempt was broadened to include, among others, the new Director, Dan Helix, and the outside consultant, Edward Burfine.

With the appearance of these two, the issue, for the first time, became a public one. As a consequence, the three engineers lost control over the matters which had preoccupied them for the whole previous year. The personal and human role of the engineers in the issue is swept aside in the ensuing conflict among management, the Board, the Bay Area Press, the California Legislature, and the various interested professional and technical societies.

Chapter 7

Management's Defense of BART:

The Firing of the Engineers

The actual operating procedures of large and complicated organizations are usually quite different from what can be deduced from tables of organization, communication flow charts and the like. Informal channels of communication are typically uncharted and unchartable. Most organizations function somewhere between the two extremes of the completely open and loose model and the closed and formal model. Accountability demands adherence to certain established and authorized communication channels. On the other hand, messages can have greater and more immediate effect if opportunities for informal communication exist. The choice of the most appropriate communication model is a very difficult one for the typical large, complex, technologically-oriented corporation.

The problem is greatly complicated in public-sector organizations where official governmental bodies (legislatures, regulatory commissions, etc.) seek to provide direct decision-affecting input to both Boards of Directors and management, and in which various formal and informal public interest and public pressure groups seek to participate in a policy-making role. The need to curb or constrain these inputs in the interest of organizational discipline and order creates dilemmas for management, and provides great temptations for professional employees seeking to reinforce their voices with the power of these groups which are technically outside the organization, but to which the organization is ultimately accountable.

Through 1970 and 1971 the character and complexion of the BART effort slowly changed. The massive construction phase of BART development was slowly coming to an end, and connections were being made among the various structural components of the system. Now, BART was beginning to install and check out the automatic train control system and to test the prototype car. The nuts and bolts direction of the project was gradually being passed from the civil and construction engineers to the electrical and electronic engineers and the computer programmers and analysts. It was at this time, too, that management felt it could, with some modest level of confidence, begin projecting a date for the actual inauguration of revenue service.

The Board, representing the public, had been applying pressure to management to set such a date for opening. Although it is always risky to project dates in enterprises as innovative and as complex as BART, it was a practical necessity to do so. If this had not been done, some of the Bay Area press would have been even more scathing in their criticism of BART management than they actually were. As a result, management began tentatively to suggest a possible opening date in late spring or early summer of 1972. From the perspective of Fall, 1971, however, the specific date projected for opening did not, for practical purposes, make that much difference. It was still not possible to distinguish confidently among, say, a June 1972, a September, 1972 and a January, 1973 date. The variables were so numerous and the possible delays so unpredictable that any date offered was not necessarily the specific one that management was really shooting for. "It was more a stimulus to create an urgency on the part of the suppliers and the engineers to achieve that date." These words of Charles O. Kramer reflect the substance of a talk he had on this subject with Assistant General Manager David Hammond.

As summer turned to fall in 1971, BART experienced a full quota of technical and managerial problems. In a sense, the system was quite as innovative in its managerial and contract-monitoring methods as it was in technical design. The best evidence suggests that nobody had ever assembled a supervisory consortium like PBTB to oversee performance on a series of contracts covering so great a variety of engineering services. But it seemed to be working. By 1971, performance on the civil engineering parts of the total project had been accomplished with a high enough degree of success that management had increased confidence in the ability of PBTB to do the job. Everyone knew that the novel Automatic Train Control System was going to be complicated, and so management was not surprised by the fact that problems arose.

Discussing the situation of summer and fall of 1971, Assistant General Manager David Hammond says: "In a project of this kind, there are lots of problems...and we were addressing a million a day, it seemed like at the time."

It was during this hectic fall that the central issue with which this book is concerned really surfaced. Word was being spread, both within and outside of the BART organization, that Westinghouse's Automatic Train Control System was having trouble and that, unless technical and organizational changes were made, the safety of the public who would eventually ride the BART trains would be imperiled.

Management mentions that it was aware that there were difficulties involved in the design and installation of the ATC System. It could hardly have been otherwise in a completely novel technology. But there was no thought of ever inaugurating revenue service until all the problems had been solved to the complete satisfaction of PBTB and of BART management.

Since problems are the very essence of trail blazing design, when questions were raised by the three BART engineers about the ATC System, they were viewed simply as part of the legion of problems routinely being addressed by management on a daily basis.

The three engineers--Holger Hjortsvang, Max Blankenzee and Robert Bruder--claim that they were forced into clandestine behavior concerning this issue because they had had no success in communicating with or getting a constructive reaction from middle or top management to the questions they had earlier raised about the Westinghouse System. In a sense, though, management can make the same claim in reverse. It claims that it had responded appropriately to written memos from them and from other members of the BART staff by reporting the issues to and requesting a response from PBTB. But when it was clear that there were some rumblings and problems at the nuts and bolts engineering level, management maintains, despite the engineers' claims to the contrary, that it was almost impossible to track them down. Therefore just as the engineers found management's responses inadequate, so did management find the source of the complaints highly elusive. David Hammond, for example, was asked: "Did you try to find out the sources of these rumblings? In a general way, were you interested in finding out what the issues were which were being raised?"

Hammond replied "Well, I'd say not only in a general way, but in a specific way. Whether they (the technical staff) knew any (problems) that we didn't already know."

In other words, if there was a communication problem, it was one which appears to have existed in both directions despite the numerous memoranda sent by the engineers to their supervisors. Still, management's desire for staff members to raise questions is a recurring theme in the story of BART management. The evidence, as a matter of fact, suggests that some of BART's higher level engineers agreed with the three engineers--to a point--that, in some areas, PBTB was not doing everything it should in monitoring the Westinghouse installation. As a result, they were not entirely unreceptive to comments of this sort from the engineers they supervised. But they wanted the comments to be presented in an appropriate and professional fashion. Charles O. Kramer, Superintendent of Power and Way, has outlined the procedure for an engineer at Hjortsvang's level to call such an issue to the attention of management.

> The route of communication was from this section or department to be carried through the chain...He (Mr. Hjortsvang) would bring this (a complaint) to Mr. Wargin's (his immediate supervisor) attention. Mr. Wargin would address that complaint to a member of Mr. Tillman's staff, who would then relay formally to PBTB our concern in a particular area.

According to management's view, the three engineers, were only three steps removed from the supervising consortium--and these steps were not through any top-level policy-maker who might be accused of hushing complaints for expedient reasons, but rather through engineers capable of understanding the technical nature of the issues and problems. Kramer notes, recall, that the complaint would not even have to go to Mr. Tillman--but only through a member of his staff to PBTB.

After explaining this channel of communication, Kramer was further questioned: "Then, under the chain of command, what would Mr. Hjortsvang be expected to do if he had a concern; (if) he relayed it to Mr. Wargin (his supervisor) and, in his professional opinion, rightly or wrongly, the problem was not remedied?"

Kramer replied: "Mr. Hjortsvang had always the opportunity to take it to a higher authority if he wished. He could talk with Mr. Ray or Mr. Hammond or Mr. Stokes."

This statement appears to accord with the policy of an open door asserted repeatedly by Stokes and Hammond and Tillman.

It is quite true that the three engineers did not all work in the same division. Hjortsvang and Blankenzee were in the Operations group, while Bruder was in Construction and Engineering, the one under E. J. Ray and the other under Erland Tillman.

Hammond

Ray	Tillman
Kramer	Fendel
Wargin	Wagner
Hjortsvang--Blankenzee	Bruder

Even so, there were plenty of opportunities for information to cross group lines. Although there was no formal, direct communication channel across from one group to the other, Kramer tells us that "an informal relationship did exist between personnel in Mr. Tillman's office and Mr. Ray's department, in the sense that there was considerable communication between the lower working levels of engineers across department lines."

If PBTB was failing in some respect to do its job of supervising BART's contractors, it was formally Mr. Tillman's group (Construction and Engineering) that was responsible for doing something about it. The Operations group (Ray--Kramer--Wargin--Blankenzee--Hjortsvang) did not have this responsibility. But in spite of this formal delegation of responsibility, weekly staff meetings were held between BART engineers and PBTB engineers which included both Operations and Construction and Engineering personnel. At these meetings, informal communications could and did pass horizontally from one group to the other--with the full knowledge and blessing of top management. It was management's belief that a certain amount of redundancy in communication opportunities between professionals was all to the advantage of the system. But obviously, no communications system works perfectly, no matter how well-intentioned or planned, and BART's system of communication was no exception.

Here, then, is how things stood in the Fall of 1971. Management was projecting dates for inauguration of revenue service--but doing this as much to prod its suppliers and contractors as to satisfy the expectations of its potential customers. The technical and economic problems were intensifying. But it was clear that the various components of the system were coalescing, and that BART really was zeroing in on the time when the system would be open for business.

During this harried period, as long as management could see that information about problems was flowing freely up and down the chain of command, across the functional division lines, and to PBTB and other contractors, it was satisfied that the communications system it had established was working. Clearly, however, top management could not itself concentrate on only one of what was literally a host of technical and managerial problems to the exclusion of all others. And management believes that nothing less than this kind of concentration would have satisfied the three dissatisfied engineers. This is, of course, understandable in view of the limited areas of technical development for which the three had any professional responsibility.

Management Hears of Discontent

When, in late Fall of 1971, management became more aware of rumblings of discontent in the organization, it invited direct comment. None was forthcoming. As far as the higher echelons could ascertain, division heads were properly forwarding memoranda from their engineers to PBTB and, based on the experience of the previous five years, PBTB had all the necessary competence to respond appropriately to these communications. Former General Manager B. R. Stokes was asked about whether PBTB was in a position to make sound engineering judgments in response to the questions raised by Hjortsvang, Blankenzee and Bruder. He replied:

> That was exactly the reason they (PBTB) were there, not only to make an objective evaluation

> there (in the matter of the questions raised by the three engineers), but to make an objective evaluation day in and day out with us as the client...You know, they had more outstanding engineers in one room than these three engineers--and (sic) three technical people represented, total. It was a matter of the job they were doing...as a consultant with a public agency for their client, and I had no question at all in this or other things that they were doing the very best professional, competent job they could.

The engineers' memoranda during the Fall of 1971 really raised no issues which were not already familiar to the PBTB engineers and to the BART technical staff. According to David Hammond:

> PBTB, or their engineers, were our primary technical arm for addressing and resolving problems. Our (BART's) technical staff was there in a supervisory job...to be sure that the right instructions were given to PBTB and also to do the things which only the owner can do to facilitate the engineer carrying out what he's been directed to do...It was a joint assessment. And, quite frankly, there was nothing new or nothing even very specific in their (the three engineers') supposed laundry list of problems.

In terms of the chain of command, the issues raised by engineer Bruder had ultimately to be carried forward by Erland Tillman, Director of Engineering and Construction. Tillman says:

> He (Bruder) had asked to see me and I had said, "Fine." And I had met with him, had listened to what he had to say, and had determined that I should look into it, which I did, and took steps to increase the attention which was being paid by management echelons in the consultant organization (PBTB) to this testing...and really felt that we were making good steps in satisfying his needs.

Tillman was further asked whether, in his judgment, there was at least some validity to the points which Bruder had been making. He replied:

> Well, at least let me say that he was in the field and if he felt this way about it, I felt that--fine--it was my job to support him and insure that if he felt additional emphasis needed to be placed on the test which he was, to a degree, responsible for, that I should do so, and so did.

Did Tillman get a prompt response from PBTB to Bruder's questions which he relayed to them? Tillman says: "I felt that I did. Yes, I felt so, and I felt that they increased their efforts in this area. Now, as matters turned out, Bruder didn't feel that they had done enough on it."

But, Tillman was questioned, did PBTB respond effectively to the questions which were raised? "In my view, they did, yes."

Now, in retrospect, it may be possible to make a more sensitive and informed judgment as to whether the points raised by the engineers were valid, central and basic to ATC, and whether the response of PBTB and BART management was sufficiently vigorous and searching. The answer to the first question is "probably yes", and to the second, "perhaps not". But from the point of view of 1971, management was apparently doing the best it could under the enormous stresses of the actual situation. And under these stressful circumstances, it appears that, management did listen, did pay attention, and did seek answers to the questions the engineers faced.

The November 18 Memorandum

A new dimension to the situation developed on November 18 of 1971. Before this time, the engineers had sent signed memoranda and test reports to their supervisors and had exchanged ideas with them and with higher management in person. But on November 18, an anonymous memorandum was circulated throughout the BART organization. It was called "BART System Engineering" and it attacked the District's design, construction and development of the Automatic Train Control system, charged a lack of technical competence in District's reliance on the judgment of PBTB and said BART was guilty of general incompetence in terms of its technical supervision of the entire project.

More than this, the memorandum went on to offer a solution to these problems. It recommended the creation of a <u>Systems Engineering Department</u> which would "coordinate designs, determine performance standards, correlate with contractors and PBTB in technical matters, and in general, assure the best possible system."

The memorandum suggested that most of the personnel for this new department could "be extracted from existing departments," but added that "a number of new specialists should be hired."

Management was faced by the difficult problem of what to do with a document of this sort. Was it written by people inside or outside the organization? Presumably inside. But who? Management asked that the author(s) come forward to present their ideas more fully, since, in fact, it had really been supposed all along that BART had been taking a systems approach to coordinating and integrating the many different activities which were being carried on in developing the Train Control System and, indeed, to managing the entire development of BART. But nobody came forward. Consequently, management reasoned that it had no alternative to assuming that, if the author wished his memorandum to be considered as a serious and constructive suggestion, he could hardly refuse to acknowledge it as his work. His failure to do so left only one interpretation. According to BART's chief administrators, the author's choice not to reveal himself indicated the existence of some motive other than simply the wish to improve the organization and its operations. The clue, according to this interpretation, lay in the sentence just quoted: "Most of the personnel can be extracted from existing departments." Management interpreted this sentence to mean that someone was using the alleged lack of coordination as a toe-hold to boost himself in the organization--to create a better job for himself. On the other hand, if management had wished to respond with greater alacrity to issues raised by the engineers, a sufficient number of written and oral recommendations had been sent by the engineers through proper channels to have warranted such a response.

Based upon this analysis--and upon management's perceptions that it was already taking a systems approach to coordination of the program, that it did pay attention to its engineers' comments about the adequacy of the Train Control System (and any other facet of the operation), and that it did communicate and interrelate on a daily basis with the PBTB specialists who were employed to monitor the contracts and supervise the technical work--management decided that it had no alternative but to ignore this anonymous memorandum, which management asserts dealt in only the vaguest generalizations with the issues of train control and safety.

For the month or so following the distribution of this anonymous memorandum, nothing more was heard. But, as management later discovered a lot was going on in secret, principally some meetings, instigated or at least abetted by a BART technician who was also a union organizer, between the three engineers and Daniel Helix of Concord, a man who had just been appointed to the BART Board of Directors a few weeks earlier.

It appears that at least two other members of the Board had been approached before Helix was contacted, but presumably when they learned of the intent of the meeting and those involved in it, they declined to have anything to do with the matter. Helix, however, denies that this was the reason for the failure of other directors to met with the engineers.

In early January of 1972, management had the first formal word that something was really amiss. It was on this date that Director Helix sent a confidential note to the President and Vice-President of the BART Board of Directors. This memorandum was basically a restatement of the materials of the anonymous November 18 memorandum. In it, Helix indicated he had met with some "second echelon" engineers who, in his opinion, represented a large majority of the District's engineers. These engineers had informed him of problems relative to Automatic Train Control and, more broadly, of BART's failure to control, coordinate and monitor the Train Control installation. Helix's prescription, like that of the November 18 memorandum, was to establish a Systems Division within the BART organization.

Two days later, a letter signed by Edward A. Burfine of Beckers, Burfine and Associates was delivered to the Board of Directors of BART. This letter informed the Directors that Beckers, Burfine and Associates had been retained by "an independent source" to "explore some of the alleged shortcomings of BART at the engineering level." The letter added that the individuals responsible for arranging for their services "had nothing to gain personally by this approach to the directors of BART" and, indeed, were "risking a great deal personally to bring to light the shortcomings and misdirections of portions of the BART system as the engineering staff view it."

The Directors were naturally concerned about this letter and they immediately asked General Manager Stokes to have his top engineering staff investigate the charges and report to the Board.

From this point on, things happened rapidly. Dissatisfied with the response from top engineering management to the charges contained in Burfine's letter and its attached report, Dan Helix somehow arranged for a series of articles to be written by reporter Rich Vogt for the Contra Costa Times during the month of January. These articles, based, it appears, on the Burfine report along with Helix's second-hand account of the issues raised by the three engineers, prompted something of a public outcry against the BART system and had the effect of seriously injuring the morale of BART's professional staff. Management views it as no accident that the articles appeared in a newspaper published by Dean Lesher, a long-time opponent of the whole BART concept, who may have seen this as an opportunity to take some shots at the system.

Finally, on January 31, Director William Reedy, Chairman of the Board's Engineering Committee, directed General Manager Stokes to respond to the technical questions raised about the ATC System and to answer a number of other questions: What were the professional qualifications of the Beckers, Burfine firm? Who requested their study? Where and when was their study made? With whom did Burfine discuss BART's ATC technology? What information was made available to Burfine? How was his report disseminated to the Board of Directors?

In view of the fact that it was a duly constituted committee of the Board--acting in behalf of the entire Board--which specifically directed the top management of BART to answer these questions, Stokes and his staff had no option but to try to obtain the requested information.

Three weeks later, the General Manager replied to Director Reedy's request. He reported that the Beckers, Burfine firm had only limited experience in the technical area relating to BART's Automatic Train Control System. Further, Stokes reported that Burfine had told members of the District's management staff that his study had been made at the request of a group of BART directors. It is clear that, in asserting this, either Burfine was misstating the facts--for assuredly, no group of directors had made this request--or else he was repeating an untruth which someone else had told him. Burfine further refused to reveal the names of BART employees with whom he had spoken concerning ATC and would not tell how he obtained technical information, other than that it had been supplied by a group of BART employees.

It turns out, of course, that in his one-day visit to BART, Burfine had spoken with no BART engineers other than Hjortsvang, Blankenzee and Bruder. All three later testified that this was so. Management's view was that one can hardly give credence to a report purporting to be an independent evaluation and body of information when the exclusive source of that information is the group which first raised the issue. No wonder that Burfine recited precisely the same complaints which management had already heard from the three engineers, and to which it had already responded after consultation with PBTB. The "majority" of BART engineers which Director Helix had spoken of had, of course, evaporated, and the dissident group was once again composed only of the three who raised the issue to start with. Throughout January and February, although the wake of the November 18 memorandum and the Burfine report were creating chaos in the organization at what was already a very difficult juncture, not one other BART engineer joined the small ranks of those who had talked to Helix and retained Burfine as a consultant.

And, according to General Manager Stokes and Assistant General Manager Hammond, even during this difficult time--largely the first three weeks of February, 1972--management repeatedly invited any employee who wished to express his opinions directly to come and see them and to do so. None came.

Management did not then--and still does not--necessarily believe that all the ideas and opinions expressed in the November 18 memorandum, the Burfine report or the various memoranda and oral communications from the three engineers to their supervisors, however broadly and generally stated, were necessarily wrong. Individual administrators at BART will agree that, in issues of this sort, there may simply be no clear right or wrong. Management simply felt that these ideas and opinions should be expressed and presented in ways which were most constructive for the whole organization as it attempted to fulfill its mission.

For example, Charles O. Kramer has testified that he did not necessarily disagree with many of the comments in these communications. But he added:

> I think they are very subjective, and they are personal opinions expressed by individuals. And I question the validity of the position expressed as to whether or not a particular way of doing a job can't as easily be done in a different way. And this is a particular direction that their background and expertise led them to believe this is the only way it could be done. And I don't share their opinion.

In management's eyes, it was the emphasis on the formation of a group actually called a Systems Engineering Division which made both the memorandum and the Burfine report suspect. A variety of views actually existed as to whether the technology of the ATC was being developed and installed in optimal fashion. But in questions of engineering design, it is almost always this way. And BART management claimed that, like most managerial groups, the were naturally most receptive to information and opinions which constructively suggested useful alternatives. But when these technical opinions were so firmly wedded to an organizational change clearly in the personal interest of those making the suggestions, it threw the suggestions themselves under a cloud of managerial suspicion. In any event, although management sought to have BART staff members communicate their ideas directly to them, this was not to be. Director Helix and the three engineers apparently preferred to go the route of the anonymous memorandum, the conspiratorial whisper in the hall, the unacknowledged sponsorship of the consultant's quickie analysis and the trial by newspaper in the Contra Costa Times.

The Actors in the Drama

In retrospect, former managers of BART have attempted, in reviewing the case, to base their judgments very specifically on the issue of who was involved in this "ethical" protest against BART management practices--and who was not. Perhaps knowing who was not is more illuminating than knowing who was. Eleven members of the Board of Directors--including all the more experienced ones--were not involved. The only one involved was a Director who had been on the Board for six weeks before he began to meet with the discontented engineers. Who else was not involved? The dozens and dozens of highly qualified engineers who were colleagues of the three discontented ones and who had been listening to their complaints. It was not that these men were not urged to cast their lot with the three engineers. Engineers D. Y. Lee, C. H. Engle, L. G. Banks, C. W. Raie and J. S. Whitely, among others, had been approached as a group by Robert Bruder and urged to appear as complainants before the Board of Directors, and Lee had been approached separately and individually. All refused. Rather, these men chose to continue to register their questions, problems and concerns through the normal channels provided by the BART organization. Of course, this refusal to join the three engineers may be as much a function of the tendency, in hierarchical systems, for employees to be more sensitive to views of persons at the top rather than the bottom as much as it is a consequence of a rational analysis of the technical issues.

But who was involved? First of all, three lower-level engineers--Holger Hjortsvang, Robert Bruder and Max Blankenzee--the last of whom had been a member of the BART staff only six months before he became involved in the anonymous criticism of the organization and of the job it was doing. Were Hjortsvang, Bruder and Blankenzee particularly outstanding experts who were peculiarly well-qualified to make the sweeping criticisms which first appeared in the November 18 memorandum and later on in other forms? Not really. All three were fairly narrow specialists doing rather specialized jobs. All three had been thought of as reasonably competent in their jobs--though Blankenzee hadn't been with BART long enough for any of his supervisors to form a really firm professional judgment--but none had been rated as outstanding or meriting promotion to a higher echelon. And these professional judgments of them were made well before they had gone to the Board member, and before the whole unfortunate firing took place.

Take one example--the previous evaluations of Mr. Hjortsvang by his supervisor. In four successive annual evaluations--1968, 1969, 1970 and 1971--on a total of 40 items of professional performance on which he could have been rated either Outstanding, Above Average, Average, Below Average or Below Standard, Hjortsvang received 13 ratings of Above Average and 27 of Average. In three of the four years, his overall rating was Competent--Average, and in one year, Above Average. His greatest strengths were stated to be his solid technical knowledge of his area and his willingness to study and improve his knowledge. His weaknesses, however, are most revealing. "Does not always see broad picture." "Lack of curiosity and initiative beyond narrow assigned area." "Tends to complain rather than suggest positive action." "Somewhat set in his ideas, resists changes in routine." "Some tendency to cling to his own concepts rather than accepting those presented by others."

This listing of strengths and weaknesses serves to show him as a good, average working engineer: competent on the job, technically sound, but limited in scope and fixed in his ways. If there is a desired model for the head of a systems engineering group in a technically novel program, Hjortsvang's personal and professional characteristics would point to him as precisely its antithesis. The systems man need not have the narrow technical competencies which Hjortsvang undoubtedly did, but he must be able to see the whole program broadly and to perceive the complex interrelationships of its parts, something which all of Hjortsvang's evaluations suggest is not one of his strengths. Even more, a systems man must be flexible and open to new ideas--to fresh perceptions of how the parts articulate to form the whole. If there is a dominant tone to the negative comments about Hjortsvang, it is that he tended to be rigid and inflexible, stubborn in his allegiance to his own views and unreceptive to the ideas of others. Of course, it is possible that Hjortsvang's views on the automatic train control system were ignored specifically because of the negative tone of some of his evaluations.

The other two engineers appear equally unlikely as individuals especially capable of understanding the problems involved in the development of the complex control system and of suggesting solutions. One of them was Robert Bruder. Bruder had been a good organization man, but never more than a methodical, plodding engineer, of whom the best Erland Tillman could say was that he was "a competent functioning employee." The other was Max Blankenzee. It is possible that Blankenzee was really an outstanding professional--but he had had only six months to observe the system. Blankenzee was basically a computer man and so new on the job that BART didn't even have an evaluation on him.

Along with these three were Dan Helix, a Director of very recent vintage and a technician-union organizer named Gil Ortiz who arranged the initial meeting between Helix and the engineers in the local union hall. This represents the totality of the "ethical" group which felt that its standards--both technical and ethical--were so high and its motivations so pure that they could impugn the professional and moral integrity of the many echelons of professional engineers who comprised BART's middle and top management and the entire PBTB supervisory structure.

This, of course, is not to say, as things turned out, that some justification was not later found for the technical position taken by the three engineers. But now we are talking about comparative professional and technical assessment of problems, and these are questions of technical judgment, not ethics. In the development of any novel technology, professionals of equally high competence and with equally high ethical standards may well take different positions. Time and experience may prove one of these positions more sound than another--but this does not justify widespread and anonymous accusations of technical incompetence and ethical deficiency.

Charging someone else with unethical behavior has all the dangers of wielding a two-edged sword. Only those who are themselves invulnerable may really feel free to do so. If, in the pursuit of some ethical ideal, individuals or groups violate another canon of ethical behavior, they seriously compromise their own position. And in management's view, this is precisely what the three engineers did. In pursuit of the admittedly ethical and benevolent goal of improving the effectiveness of the train control system and thereby protecting the public safety, the engineers cast aspersions on the professional integrity of the PBTB engineers and the supervising engineers in the BART organization, undercut morale throughout the organization and, then, to compound this behavior, failed to own up to their actions but instead flatly denied their roles in the incident. Under these circumstances, it was and is difficult for management to take their ethical stance very seriously.

The Crisis Mounts

At a February 22 meeting, the engineering committee of BART's Board of Directors rejected the Burfine report. This committee rejection was sustained at a February 25 public meeting of the Board of Directors which was attended and addressed by Burfine. The vote of the Directors to support BART management and its administration of the project was 10-2.

The Directors' judgment concerning the quality of the Burfine Report paralleled that of management. David Hammond, speaking of the Burfine Report, later said:

> You know it was a one-day report, which takes quite a bit of genius to do in something as complicated as this...I don't think it was very professional. It wasn't a professional thing to undertake...It was superficial, and it couldn't be anything else.

Similarly, B. R. Stokes has characterized the Burfine Report in the following words:

> My first reactions to the report was that it was a ridiculously unprofessional, incomplete...almost silly effort in terms of anything smacking of professional engineering.

The really interesting thing about the Burfine Report is that the negative opinion of it expressed by management and the Directors was later supported by at least one of the three engineers. Eighteen months after the incident, Holger Hjortsvang, writing to Professor Stephen Unger of Columbia University (July 21, 1973), states:

> Mr. Burfine tried to elbow his way in as a BART consultant, following a conversation with Blankenzee and me. He wrote an un-called for, critical but entirely incompetent report which led Helix to the fatal decision of asking him to present his views to the Board of Directors. He was an easy target for the specialists from Westinghouse and PBTB. Burfine was certainly not engaged by me or the two others.

The appalling and (if the problems BART was facing at the time had not been so grave) almost amusing finale to this episode was that Burfine asked the BART Board to pay for his work in preparing the report. He had apparently been told by someone that it was an effort which the Directors had authorized. As far as management knows, nobody connected with BART ever paid Burfine for his efforts.

In the meantime, since nobody had come forward to acknowledge responsibility for writing and circulating the November 18 memo, for communicating with Director Helix and with the *Contra Costa Times* or for engaging the services of Mr. Burfine, management simply took note of the limited and questionable technical points which appeared in these documents and consulted with its PBTB monitors concerning them. In addition, following the directive of the Board, management attempted to find out which of its staff had prompted or participated in these activities. But the attempt was pretty low key.

It was low key because it appears that none of BART's section heads could really believe that any of their staff engineers could have been guilty of such unprofessional conduct.

B. R. Stokes later recollects the directive of the Board concerning identification of the dissatisfied staff members and his own action in response to it:

"I do recall asking my senior people to attempt to find out where they (the various *subrosa* communications) were coming from and identify those who were responsible."

In view of the strong and insistent accusations against BART management which had been and were being made, this does not, on the face of it, appear to have been an unreasonable or unprofessional action.

The actual request to identify the BART employee(s) who had been in touch with Helix and Burfine--and perhaps the *Contra Costa Times* --was so muted that a number of top managers don't have a clear recollection of it as a directive at all. Erland Tillman, for example, was asked whether he and his senior staff were charged by Stokes and Hammond with trying to identify these people. He replied:

> Oh, I think I would have to say "No" to that. I don't remember anybody asking me to endeavor to find out who they were. And...I don't know...don't remember how I first learned that Hjortsvang and Blankenzee were the two people. I just don't remember how they were identified...

> Oh, I think I would have to say "No" to that. I don't remember anybody asking me to endeavor to find out who they were. And...I don't know...don't remember how I first learned that Hjortsvang and Blankenzee were the two people. I just don't remember how they were identified...

> Well, somewhere along the line we got the impression that it might have been Bruder and so I, as I remember it, I asked Fendel to try to find out, and he, to the best of my knowledge, came back and said, "No, he didn't think it was", and then I was led to believe that it might be Bruder, and so I asked Bruder and Bruder said, "No, it wasn't him...he had never had anything to do with it."

> Shortly after that, Wagner came to me and said to me that Bruder had told him--they happened to be riding home together--that he had participated in a telephone conversation with Blankenzee--no, it wasn't Blankenzee, it was Hjortsvang--and Dan Helix. And...the dates were such that this telephone conversation had

> taken place before Bruder had told me that he had nothing to do with it.
>
> Well, Wagner told Fendel and then Fendel told him to come in and see me, and Wagner personally told me.

According to this account, then, the discovery of the identity of the three engineers was partly directed and partly fortuitous, but apparently not the result of any managerially dictated inquisition.

C. O. Kramer's later independent recollection reinforces Tillman's. Kramer was asked whether, after he had gone over some technical aspects of the Burfine Report, he discussed with Mr. Burfine the source of his information.

> Yes. Mr. Burfine did not relate to us the source of this information.
>
> You asked who gave him the information?
>
> Yes
>
> And he said he wouldn't tell you?
>
> That's right...
>
> Did you tell Mr. Hammond that Mr. Burfine would not reveal the identity of his sources?
>
> Yes.
>
> Did Mr. Hammond or anyone else ask you to make further efforts to ascertain their identities?
>
> No. They did not and I did not make such an effort.
>
> You never made any effort to find out who the men were who talked to Helix or Burfine?
>
> No.

> Never questioned any of the engineers?
>
> No.
>
> Did you ever ask Mr. Wargin to question any of the engineers?
>
> I did not specifically direct Mr. Wargin to question any of the engineers.
>
> I did ask him who he thought was the originator or instigator of the material that was furnished to Mr. Burfine, that is the basis for this report...
>
> What did Mr. Wargin respond?
>
> He said at that time that he didn't think that it was anybody at BART. I didn't either.

Top management was naturally anxious to know who had been meeting with Helix and Burfine, and asked if anyone knew. But beyond this normal and expected interest in discovering the identity of the BART employees, there appears to have been no organized hunt, echelon by echelon, for the engineers, and independent testimony provided by Stokes, Tillman, Ray, and Kramer is consistent on this point.

We find ourselves, then, at this juncture. On February 25, 1972, the Board of Directors, after reading the Burfine Report and hearing from Mr. Burfine in person, voted 10-2 in effect to reject his report by giving a vote of confidence to General Manager Stokes and his management team. At the same time, management feels it had not ignored whatever hints of possibly useful material may have existed in the report, in the celebrated November 18 memorandum, in the stories appearing in the press or in any of the in-house memoranda submitted by Hjortsvang, Blankenzee and Bruder and any other BART engineers. It had been pursuing with PBTB and with Westinghouse some of the questions which had been raised, and had been requesting responses specifically about the documentation and testing of the Automatic Train Control System. Also, at the same time, management was virtually imploring members of the professional staff who harbored doubts or reservations about the technology of ATC to come forward and discuss their feelings at any level they chose up to the general manager. Mark Bowers of the BART Personnel Office has written in a file memorandum that Hjortsvang and Blankenzee specifically (since they had written memos on the subjects) "were asked to work with us on this problem three and four times respectively by Director of Operations E. J. Ray. These engineers repeatedly denied any knowledge of the situation involving Burfine or Helix..."

The Burfine Report

Of some interest is the fact that, although Dan Helix and others interested in attacking the management of BART have relied very heavily on the Burfine report to support their views, even at this early date, engineer Hjortsvang anticipated his later contempt for both Burfine and his report. He has testified that he was never impressed with the quality of Burfine's analysis and indeed, suggests that the Burfine Report was generated and developed almost by accident--with nobody assuming the responsibility for having commissioned it--and Burfine, one may infer, pushing it because of his interest in collecting a consultant's fee. Hjortsvang has testified that the only motivation to get a consultant was to reinforce the engineers' case before the Board. Hjortsvang reported that, in conversation among the engineers, Blankenzee said, "I think I know somebody" who would create more credibility with the Board "to sort of reinforce what we said."

Hjortsvang later added that Burfine was not even asked to come and inspect the BART facilities but "came along to get oriented--

that's what I thought, at least--oriented about the meeting (with the Directors) and what we wanted him to assist us with."

Burfine showed up at the BART office but, according to Hjortsvang, was not asking to "go into any specifics about BART." Instead, he was simply to be available to talk about computer control systems.

> Nevertheless, (Hjortsvang went on) "he started asking questions about the system, to my consternation, because I didn't want to discuss BART's internal things with him at all...And I told him several times when he questioned these things that I didn't think they were of any interest in connection with this meeting we were going to have. But I...I could not stop him, mostly because I think Blankenzee felt that he should at least accommodate him and answer his questions and tell him about what we felt was wrong.

Hjortsvang says that Burfine was on BART premises something over half-a-day and that he (Hjortsvang) and Blankenzee answered his questions. But then Hjortsvang was asked:

"At that time did you have in mind that Mr. Burfine would make a written report?"

"No, definitely not."

Burfine, for his part, later claimed that Blankenzee told him that he (Blankenzee) wanted a short study of the ATC System, that "the report would be turned over directly to the Board of Directors, and that a group of Directors, through him, were seeking the report."

Blankenzee, according to Burfine, indicated "there were two other engineers involved...and that there were at least two Directors, namely Gil Ortiz (sic) and Mr. Blake."

Further, Burfine states that "He (Blankenzee) indicated that this group of Directors would fund the study."

Burfine later claimed that he gave Blankenzee a quotation for two days' consulting plus a secretarial fee, plus some other miscellaneous expenses that would be the price for his report. Blankenzee allegedly said, "Okay, I'll have to check if that is acceptable." He is then supposed to have called Burfine back to tell him the fee was acceptable and that he should proceed on the project.

It appears that everybody associated with retaining Burfine--Blankenzee, Hjortsvang, Helix--later disclaimed any knowledge of a commitment to pay a fee and even, in the case of Hjortsvang, won't even acknowledge that Burfine was retained to do a report at all. Naturally, that the Directors later refused to have anything at all to do with Burfine's report--to say nothing of paying for it--is hardly surprising.

The Ethical Claim

The claim is made throughout by the three engineers that they were motivated in going through this series of actions by a basically ethical concern for the future users of the BART system, and that their (from management's point of view) unethical behavior--circumventing organizational channels, making false implications to Burfine, undercutting organizational morale by trying to set engineer against engineer within the system and, finally, lying to the very end by denying that they had anything to do with the whole business--was justified by the ethical end for which they strove. Management responds that, although the argument that an ethical end justifies use of unethical means has been sufficiently demolished time and time again, in practice as well as theory, a little respectability may be lent to what was undeniably a devious pattern of behavior by purity of motive. But management, by probing a little more deeply, believes that even the purity of the end must be most seriously questioned.

Why, in a word, did the three engineers launch this whole enterprise?

Of course, three different people are involved--each with his own personal attitudes, motivations and goals. It is difficult enough for each of us to discern clearly his own reasons for doing the things he does, and it may be presumptuous for anybody to try to tease out the basic motivation for the actions of others. Still, the question is critical enough that the attempt should be made. Fortunately, there is some documentation available to assist the analyst.

The three engineers were at quite different stages in their careers when they launched their attack on the ATC System and on BART management. Mr. Hjortsvang was near 60 and had just a few professional years ahead of him. Mr. Bruder was younger--around 50--and should have been in the prime years of his professional career. Mr. Blankenzee was in his thirties and was still preparing himself and developing professionally. These differences alone might suggest some differences in motivation.

Another difference exists in the engineers' areas of professional specialization. Hjortsvang and Blankenzee worked together in the technical area involved in the ATC System, and Blankenzee specifically knew this aspect of computer technology. Bruder, however, had little to do with Hjortsvang and Blankenzee on a daily basis, and his responsibility was not nearly so closely related as theirs to the issue of automatic train control and associated questions of safety and reliability.

It appears that the leadership in raising the issue came from Hjortsvang and Blankenzee. Hjortsvang had been with BART since 1966, and had established a reputation as a competent--if somewhat rigid--engineer. Blankenzee did not join the organization until May, 1971. Hjortsvang had apparently questioned some of the ATC planning early in 1971, but it was not until after Blankenzee joined BART that the questioning became really insistent. Even so, the memoranda written to their supervisors by

Hjortsvang and Blankenzee had been perfectly appropriate professional communications, and had represented the kind of independent thinking which BART management had tried to cultivate in their own employees over the years. Management spokesmen insist that it was a prime BART management principle that only when issues and concerns were fully aired could they adequately identify--and correct--problem areas.

But as far as management was concerned, the anonymous November 18 memorandum represented a real departure. This was no longer a professional communication designed to call the attention of senior engineers and managers to problems in the system. Instead, in its wording and its indiscriminate circulation, it was clearly planned and designed as a broad attack on the competence of BART management and on the professional abilities of the PBTB supervisory engineers. Management felt it could not construe a sweeping attack of this sort as a professional communication; it had to interpret it as a political document, designed, not to identify problems and propose engineering solutions, but rather to start a bandwagon, recruit support and force structural changes by popular pressure. The changes desired were clearly stated in the memoranda's last paragraph: "...it is proposed to establish a Systems Engineering Department immediately. Most of the personal can be extracted from existing departments..."

Hjortsvang, the author of this memorandum, was well aware that, in the structure of BART, he was the engineer whose seniority, whose field of professional expertise and whose current assignment came the closest to the vague and general area which has come to be called systems engineering. If the leadership of such a new department were, in his words, "to be extracted from existing departments," the great probability was that, as the senior person in the area, he would be chosen.

This attitude, sensed by BART management in early 1972, was later clearly and specifically confirmed in a letter written on February 5, 1973, by Hjortsvang to Helix. Hjortsvang, discharged from BART a full year before writing this letter, still envisioned himself in a position of leadership, authority and power. He wrote to Helix:

> My own opinion is that to comply faithfully with the directions indicated by the investigations, BART needs a definitive and competent technical staff, headed by a manager who can generate the lost confidence of the Board, the Legislators, the Public Utilities Commission, and the public in general.
>
> At this time, I believe, it would be difficult to find an independent engineer besides <u>me</u> that could generate this confidence by all parties. This is why I take this opportunity to appeal to you for your help in convincing the Board of my suitability for a new post which could be called Board Nominated Manager for Systems Engineering and Transportation.
>
> Senator Alquist (Chairman of the Senate Committee on Public Utilities and Corporations) says in his statement today: 'Quick conformance with these recommendations is absolutely essential if faith in the system is to be maintained.'
>
> I have plans ready for the structure of the new Engineering Department, and for the personnel needed.
>
> I also have developed outlines of the necessary design improvements, and of the realizable performance criteria.
>
> In other words, I am ready to go to work immediately, with no need for a break-in period to get acquainted with the system and the problems.
>
> P.S. An amendment to the BART law may be required, but I suppose this would be a small

> matter for Senator Nejedly (a member of the Senate Committee on Public Utilities and Corporations)."

A clearer demonstration of Hjortsvang's motives during the Fall and Winter of 1971-72 could hardly be made. To confirm this view, we have the words of another of the three engineers, Robert Bruder. Bruder was asked:

> It is clear in some of Stokes' statements...that he felt that you and Blankenzee and Hjortsvang were motivated by self-interest...That you were trying to set up, the three of you now, trying to set up a Systems Engineering Group in which you would hold high position--essentially a power play within the organization.

To this Bruder responds:

> The prime leadership in this whole thing was Blankenzee and Hjortsvang...and I think on Hjortsvang's part in particular, there was that motivation.
>
> It was so obvious when I met the group and it got more so as we worked. I have no doubt. There's no contest there--no contest. The guy is right...
>
> He had been through it all and he did have some motivation. He was out for Holger (Hjortsvang).

Bruder does not, however, perceive self-interest as the motivation for Blankenzee's participation in the episode. Rather, he expresses admiration for Blankenzee's high principles and professional competence. On the other hand, the testimony of other BART engineers, specifically that of Robert Fickes, tends to suggest that Blankenzee did indeed see the issue as a power struggle between adversaries, one of whom would end up in a position of authority. Fickes retrospectively recalls that Blankenzee told him that E. J. Ray shared his (Blankenzee's) opinion that a Systems Engineering Group should be formed to reserve BART's engineering problems.

> Mr. Blankenzee (recalls Fickes) indicated that he would be the most qualified to fill the Systems Engineering slot because he would make a better supervisor than Holger Hjortsvang, Blankenzee's supervisor...After one of his alleged meetings with Mr. Ray, about three or more weeks prior to the dismissal of Hjortsvang and Blankenzee, Mr. Blankenzee advised me in the presence of other fellow employees that I would have to decide which side I was on.

The sides Blankenzee was referring to were apparently existing BART management, and specifically Mr. E. F. Wargin and Mr. C. O. Kramer, and the new "Systems Engineering Group" that was to be managed by Blankenzee and Hjortsvang. Apparently encouraged by Daniel Helix in his belief that he and his allies could convince the Board of the correctness of their position and persuade them to establish a Systems Engineering Department, Blankenzee spent much time talking to fellow technicians and engineers at BART, putting them on notice that they would have to take sides in the coming power struggle and urged them to support him along with Hjortsvang and Bruder. It is illuminating to see that, despite any words of sympathy and support which some of the BART engineers may have offered Blankenzee, none said or did anything publicly which would suggest any agreement with either the technical objections or the political objectives of Hjortsvang, Blankenzee and Bruder.

Bruder's motivations are less clear. It appears that Bruder did have a strong professional conscience and that he was convinced by Hjortsvang and Blankenzee that there were serious defects in the design and testing of the control system. This system, however, was somewhat tangential to Bruder's area of responsibility, and he appears simply to have taken the claims of the other two on faith and have gone along with them. In his case, an apparently sincere but, in management's view, misguided allegiance to professional ethical behavior seems to have been an important motivation to action.

Management Acts

To return to the chronology, following the meeting of February 25, at which the Board overwhelmingly expressed its support of BART management, it became increasingly clear that Hjortsvang, Blankenzee and Bruder were the employees responsible for the November 18 memorandum, for meeting with Ortiz and Helix, for the Burfine report and for the series of denunciatory articles which were appearing in the _Contra Costa Times_. Bruder's mention to his supervisor of a meeting with Helix, the similarities in language of the November 18 memorandum to language which Hjortsvang and Blankenzee used in attempting to recruit support from among their colleagues made it clear to management that they were the three employees whose actions were having such a devastating effect on staff morale and efficiency. In the words of E. J. Ray:

> It was causing a lot of dissension...Work wasn't getting done. It was causing a lot of disruption, and I was trying to find out what the problem was and to get it settled and get back to work.
>
> It wasn't that I was particularly trying to find out who they were. I was trying to find out who they were so I could talk to them and get their complaints myself.

This was the dominant tone: if there were people this upset about technical issues, management wanted them to speak up and let it know in detail what their perceptions of the problems were, rather than to resort to the very general and nebulous sort of criticisms which were made in the November 18 memorandum and the Burfine Report. A number of people were asked whether they were involved in the episodes. All--including Hjortsvang, Blankenzee and Bruder--denied any involvement. The weight of evidence, however, became simply overwhelming. Finally, one last time, each of the engineers was asked point blank whether he had had anything to do with Helix or Burfine. Management asserts that each continued to deny any involvement. Under the circumstances, management felt it had no choice but to ask for their resignations or to fire them.

Erland Tillman recalls the last four days of Robert Bruder's employment.

> I had said to Bruder, OK. Level with me. Tell me what the score is...I have to say that I was having a great deal of difficulty really pinning Bruder down. He seemed somewhat incoherent about the whole thing. Apparently he felt himself under...under a fair amount of stress, as I look at it...a very few days after that that I got the word from Wagner that Bruder had...was, in fact, a part of the group that had been...was involved in the Burfine report and was involved in direct access with the Board rather than...than with the supervisors on the staff.
>
> Question: Oh, at that point, really the notion of firing the members of that group was being entertained by management...?
>
> Tillman: At least the notion of firing Bruder was being entertained by me, I'll say that. Strictly on the basis of...you know...well, let's face it...of lying to me. You know, I'm not used to having people tell me one thing while they're doing something else.
>
> Question: Yes. Did you make the recommendation to Hammond and Stokes that Bruder be fired?
>
> Tillman: Ah...in Bruder's case, I sure did.
>
> Question: And I gather that...that Hammond and Stokes accepted your recommendation without reservation?

Tillman: Well, Stokes required me to do a fair amount of checking before he would agree. And one of the things which he required me to do...somebody had got the...the word that Bruder had talked to a number of the people in the operations division and had requested them...or asked them whether or not they would be willing to appear before the Board, and...testify that they weren't satisfied with the testing that was going on, and so on. And Stokes asked me to talk to these people, which I did, and I found that Bruder had said something to a group of them, I guess about 30 or more of them, along that line, and then, then, when that group broke up, had followed one of them into his office and again had talked to him about, would he be willing to...at least the answers they gave me was, that the only basis on which they would talk to the court would be direct at the request of management or...or under subpoena. And later on when I...when I raised the fact that Bruder was at, which I did at the...our last interview, he said to me that that wasn't the case at all, he had...he had merely...they had been having a bull session and he had just kind of thrown that out in conversation, and when I said, "Well, that might be for the big group, but when you follow one of the individuals into his office and again said it to him, that doesn't sound to me like objective conversation." But we never...we never reached any...any meeting of the minds with Bruder on that.

Question: I see...well, let me...I...I'll ask you to conjecture. Maybe you'll choose not to. If, at the outset, when you asked Bruder whether he was involved in this, if he had not lied, if he had simply said, "Yes", do you think the outcome would have been any different?

Tillman: Well, I think that I can honestly say that my attitude toward Bruder...would have been entirely different. Because my attitude under those circumstances would have been to...to kind of counsel him, and say, "Look, this...this isn't the way to go. You shouldn't do that. I think you're wrong on this." Now whether or not it would have made any difference, would have kind of depended upon what his...his reaction would be. But I...I feel

> quite sure, even after all this time, that my attitude and my reaction would have been...would have been totally different than it was.

B. R. Stokes' recollections of the motivations and the actions of the engineers is similar. He recollects the last few days like this:

> At this point, some of the true motives were beginning to come out that it was not so much that anything was wrong with what was going on at the moment, but it could be done better with a...with a...I believe it was called a Systems Engineering Group...three of them would be very very highly involved at a very high level...At this point there were comments being made to various members of the same middle and lower level technical staff, that "You'd better begin to choose up sides, we're gonna take over and if you want to go along with us, you"re gonna be taken care of." And I think it was this, more than anything else, that identified at least two of the people involved.
>
> I do recall that our people began to get reports that so-and-so had been approached and saying that "either you're with us or against us, and if you're against us, you won't have any new place in the new organization when we take over."
>
> Question: Do you remember who made the specific recommendation to fire the three engineers, or was this just like a consensus of the top staff at BART?
>
> Stokes: Well, I specifically authorized it, and directed it after satisfying myself that we had no other recourse, after repeated exhortations for people to come in and talk, after the material continued to flow into the press, and after continued frustration at getting the...the havoc that was being caused among the staff cooled off. So when I found myself with no other choice, I authorized and directed.
>
> The recommendations from their specific department heads followed several efforts

to...to...you know, to get the truth out. And faced with continued and continual lying and continuing efforts on their part to disrupt staff morale and disrupt the orderly processes which we were trying to carry out, on recommendation from the specific department heads, I said, "Fine, let's go."

Question: OK. From the time when the three were identified, did you ever personally sit down and talk with them about their behavior or about the technical issues?

Stokes: Well, yes, I specifically talked with Bruder and arranged for him to, that same day, I believe, to have a conference with Mr. Hammond and Mr. Tillman and others, I believe, and that was followed up by a further and more detailed conference. And Mr. Bruder again was still lying at the time, and apparently, as I recall, and again with difficulties, but he declined to produce certain back-up documents and information which he was asked to produce. And it was only sometime after that that--and when I say sometime, I'm not sure whether it was a day or a few days, he was, in fact, identified as one of those who had been engaged in these activities, and he was, then, on the recommendation of Tillman, fired along with the other two.

Question: So there was no attempt--I'm using maybe an inappropriate word, to talk to and counsel with and "rehabilitate" the engineers _after_ you knew who they were. You felt that the damage had been sufficient that you had no other recourse but to fire them?

Stokes: Again, this was largely a matter for their individual supervisors and superiors reporting to me. In the case of Hjortsvang, I think, yes, if my recollection is correct, I did try to counsel with him and ask him what recourse I had, and I believe I was told in the meeting that he said, "You know, you have no other choice but to fire me."

Finally, David Hammond recollects the last few days like this:

Question: You felt you didn't have great success in ferretting out the hard specific issues?

Hammond: Well, we didn't because, this has always seemed peculiar to me, frankly that, these guys can become Sacco and Vanzetti's, denied their first amendment rights to speak out, but they kept denying that they were the ones that were saying anything. It was suspected by our middle management people who they were--they were asked and,--no, they didn't have anything to do with anything.

Question: Let me jump ahead and just interject a question. This has got to be pure conjecture on your part. If, early on, the engineers had admitted that it was they who had raised the questions and they who had written the November 18 memorandum and they who had seen Board members, do you think that management action would have been the same as it turned out to be?

Hammond: Well, I don't think the ultimate action would have been the same, because, as I said, our whole style of management was to hear from our staff as to what are the problems and what should we be doing about them. So we not only didn't say, go away and don't tell me about any problems, that was the whole purpose for the existence of anybody on the staff. To recognize problems and be addressing them.

He (Bruder) did come in (to Tillman) with a list of problems that he had--they really didn't relate much to the safety or the reliability of the Train Control System--they related more to what would ordinarily be in his area of, getting the installation completed and checked out as close to the scheduled time as possible. And great attention was payed to complaints that he made. He denied, however, that he was a part of the group that had been known to have gone to a couple of Board members, but not definitely known as to who they were.

Till the very end both Hjortsvang and Bruder denied that they had any part of it; and so did

> Blankenzee initially. But I guess he later bragged to a reporter that he was one of them. So the real basis, whether they recognize it or not, goes back to again something I said earlier--how can your first amendment rights to speak out be denied if you deny that you're speaking out? And we certainly didn't have any great confidence in people who were trying to be around sowing this kind of seeds of dissent and not willing to say, "yes they were."
>
> Question: How was the decision to fire them made? Was it...the decision of Mr. Stokes, was it a recommendation of the engineers' supervisors, or was it a joint matter?
>
> Hammond: Well, it was pretty much joint, but I would have to say that I think I was the one that primarily brought it to a head, on, to me, the very valid grounds that I've just discussed--that you can't tolerate in an organization people who don't act either responsibly or professionally.

Finally, on March 2, E. J. Ray and C. O. Kramer were authorized to ask Hjortsvang and Blankenzee for their resignations. If the two refused to resign, Ray and Kramer were authorized by Stokes to fire them. The reasons given were clear and compelling.

1) Repeated insubordination.

2) Falsification of District information and lying to their superiors and to members of the Board.

3) Perpetrating actions creating severe staff disruptions and seriously impairing staff morale.

4) Failure to perform the jobs for which they were employed in a competent and acceptable manner.

5) Repeated refusal to follow understood procedures for bringing job related problems to the attention of District management.

Hjortsvang finally agreed to resign when he understood what financial benefits were due him, but Blankenzee refused to resign and was told that he was thereupon fired. On March 3, Erland Tillman, in the presence of John Fendel, made the same request of Robert Bruder, who likewise refused to resign. Tillman, therefore, told him that his employment with BART was ended.

Fendel recalls that, when he and Bruder drove back to their office together after the final meeting with Tillman, he mulled over Bruder's comments to Tillman:

> There are some things that a man has to do. I did only what I considered to be necessary and proper.

And Bruder added, speaking to Tillman of management's action in firing him:

> And I understand that you have to do what you think is proper.

And that, in the most eloquent possible words, explains the action of the management of BART.

Chapter 8

Division on the Board:

Management Is Attacked and Defended

The directors of large organizations are often as diverse in their backgrounds, interests, goals, and values as the members of management with whom they must share authority. In some cases, board members have strong ties with, and obligations to, constituencies outside of the organization. Directors can represent a financial institution that has extended a large loan to the organization; they can represent another organization that is either a supplier or buyer of materials or products; they can be elected representatives of political units such as cities or counties; or they may represent public interest or consumer groups. In some cases, board members are also members of the management of the organization, thereby producing strong internal coalitions between the board and management.

The power of a board of directors relative to internal management depends heavily on their ability to have an independent source of information with which to evaluate the actions and recommendations of management. This normally requires a staff of persons with appropriate expertise.

In the absence of a staff and independent information, directors are often powerless in the face of management actions. In such cases, political coalitions may determine the nature of the relationship between the board and management.

In October, 1971, Daniel Helix was appointed to the BART Board of Directors by the Contra Costa

County Mayor's conference. With this appointment began six months of stormy controversy between Helix and BART management over the actions of engineers Hjortsvang, Bruder and Blankenzee, which culminated in their being fired in early March of 1972.

Helix had campaigned actively for the board position, an almost unprecedented action. As Helix recalls it, he was motivated to seek the seat on the BART board because, as city councilman, vice-mayor and mayor of Concord, he had become convinced that BART was discriminating against the outlying towns and in favor of the heart of the metropolis. For example, he had heard the complaints of the Concord City Manager that BART management would not communicate or cooperate with the local Park District concerning the aesthetics of Concord's BART station. The station, a stark gray concrete structure, was the last BART stop in Contra Costa County, and Helix was convinced that it compared most unfavorably with the elaborately designed, muralled and mosaicked terminals in San Francisco, Oakland and Berkeley. In an earlier effort to facilitate communication, Helix had tried to call B. R. Stokes, BART's General Manager, about the problem. After receiving no answer, he tried to reach the former General Manager, John Pierce. Again, he got no response. As a consequence, Helix concluded that effective communication could come only through his appointment to the Board, where he could wield clout along with conviction. And so he began his campaign with the Mayor's conference of Contra Costa County, a campaign which was rewarded by his appointment to the BART Board in October.

Within five short months, Helix had a major impact on the Board, on management, and on the situation of the three engineers. So active was Helix, as a matter of fact, that the story of the Board of Directors from October, 1971, until early March when the engineers were fired is essentially the story of Dan Helix.

Daniel Helix, a handsome, loquacious and likable man, was associated in early 1971 with the Transamerica Title Company. Immediately before joining Transamerica, he had been administrative assistant to California State Senator Nick Petras. As he recalls it now, it was serving in that role in Sacramento "...that whetted my appetite...for the political life and I saw some things while working at the state level that needed to be done locally. I decided to run for office (Councilman) in the city of Concord in 1968 and I was elected..." From 1968 on, therefore, Helix held two jobs, one with Transamerica Title, and the other in local politics. After serving later as vice-mayor and mayor of Concord, Helix's political career ended after a primary defeat in the race for Assemblyman in the State Legislature. By 1977, Dan Helix had left Transamerica and was operating the Walnut Creek, California office of Chicago Title Company, an office he built and which he proudly states he took from dead last to fourth of nine companies in market share in just two years.

Today, Helix's office walls proclaim his public involvement. On them hang Resolutions of Appreciation from the California Senate, citations for his work as Mayor, and citations for other civic service, as well as a proclamation expressing appreciation for his service on the BART Board. Although he is now out of public life, Helix is still active and still proud of his years of public involvement. In retrospect, Helix recalls the large share of his time and energy which public activities took, but still expresses great commitment to the public's welfare and to the need for better government and more involved people. He is a man who believes deeply in the democratic process, and his views of what democracy is and should be are clearly reflected in his liberal viewpoints and Democrat party affiliation.

Helix Joins the BART Board

Helix began his tenure on the BART Board in October, 1971, with a basic credo that BART was owned by the people. But the people, he feared, like the cuckolded husband, are always the last to know. In the case of BART, what they did not know was how their money was being spent. With such an orientation, it is not surprising that Helix would soon see much to confirm his fears.

Helix, however, recalls a rather benign start to his BART board service:

> My first meeting in November was rather routine...and I expected that my role on the BART Board would deal with things like the collection of fares, rate setting, some union activities with employers and things pretty much along the same lines as I'd been experiencing as a Council Member in Concord.
>
> I didn't even expect to get into the bag of worms that I subsequently got into.
>
> And that (routine) was fine with me because primarily I wanted to make sure Concord would not continue to get what I felt was the short end of the stick in terms of the monies being contributed to the stations and things like it. If you start in a central area the BART stations are ornate. You have special imported tile from Germany, you have all sorts of architectural designs that are unique and unusual, and then the stations get progressively drearier as you get out along the system, and, of course, Concord is the terminus and so I was apprehensive about the steel gray concrete walls which I thought we would and finally did get.

If November was routine, December saw things starting to heat up. As Helix puts it:

> We (the Board) received what was to be a routine report from the General Manager that we had two trains on the system at Oakland and they happened to have found one another. One was stationary and the other was moving on a practice run.
>
> There was an investigation and the Board was advised that the conductor driving the train

> had fallen asleep and so it was pilot error and the matter was closed.
>
> Well, there had been some comments in a local newspaper, I believe it was the (Oakland) Tribune, but I'm not positive, believe it was the (Oakland) Tribune, but I'm not positive, that the conductor when interviewed after the accident stated that he couldn't stop the train.
>
> Mr. Stokes advised us (the Board of Directors) that a Board of Inquiry had been convened by him, without my knowledge, or the knowledge of any other member of the board that I talked to.
>
> The (investigating) Board had been dismissed. They had done their job. They reported pilot error. So I said, 'Was the question of the inability of the train to stop because of some automatic malfunction presented to the Board of Inquiry?' The answer (from Stokes) was, "No."
>
> I said, "I find that surprising and I'm going to move that the Board of Inquiry be reconvened to investigate the charges that there is a mechanical malfunction in the equipment"
>
> Mr. Stokes was upset about this and he had a way of letting me know when he was upset. He would frown and look down at the table and move his head from side to side and then sit back and puff on his pipe and say, "Well, I think you are absolutely wrong."
>
> Anyway, the motion passed, barely, but it passed. So I said, "Fine, that's all I'm interested in, simply reconvene this Board of Inquiry." Well, I might tell you that to my knowledge (the Board of Inquiry) hasn't been reconvened as of today.

Helix later added that he had never seen a written report on the accident even though one was supposed to have been prepared. This direct violation of the Board's motion was very surprising to Helix. He reports having spoken to Stokes, after learning that the Board of Inquiry had not been reconvened, he said:

> It's passing strange that you can have an accident and you can have a Board out here two days later and dismiss them two days after that and then get directions from the Board (of Directors) to convene them and now three months later we still don't have information.

Helix used the incident to illustrate the support Stokes had on the Board of Directors.

> Then this little drama would take place. One of the directors would say, "What's the matter, the general manager's got a lot of things to do. He's trying to get a railroad into operation. Why don't you get off his back?" This at a Board meeting.
>
> And somebody else would say, "Mr. Stokes do you have that in process?" And he would say, "Oh yes, I have it in process." "Fine, let's drop the matter and go on to something more important." So the matter would be tabled and it was useless to fight against the gavel.

Helix Meets the Three Engineers

The December 1971 accident came on the heels of the anonymous November 18 memorandum which was widely circulated among BART's employees. Helix believes it was his inquiries concerning the December accident which drew Bruder, Blankenzee, and Hjortsvang, the author of the memorandum, to him. Reports vary, however, as to exactly how the contact was made. Gil Ortiz, a labor organizer, and a technician at BART, was said to have been the middleman in arranging a meeting between the engineers and Helix. As Helix remembers it:

> I got a phone call from a person telling me that he was an engineer and that he wanted to talk to me about that accident. And would I meet with him at the union office?
>
> I asked, "Why don't you talk to the General Manager?" He said, "We tried to talk to the General Manager and we can't get to him." I said, "Yes, that I'd meet with them."

The engineer who called was Max Blankenzee, according to Helix, but Helix suggested that Ortiz did act as a go between. When asked why Ortiz contacted him, Helix replied that it was because of his prior association with Senator Petras, who had always had labor support. The meeting took place at a labor hall near Lake Merritt in Oakland and BART Headquarters.

Ortiz's role was to provide a meeting room and to introduce Helix to the three engineers. After the introduction Ortiz retired to his own office and participated no further in the meeting.

Helix reports:

> We sat down and then this material began pouring out. They gave me a copy of the (Burfine) report. They told me that two directors had agreed to pay for the report.

Out of BART funds or out of their own pockets? "Well, I don't know how," replied Helix. It was Helix's understanding that, "...the engineers had paid for the report and they were going to be compensated by two directors, Bill Blake and Nello Bianco."

The Burfine report was prepared by Edward Burfine, a consulting engineer connected with Beckers, Burfine and Associates. Just how the engineers sought and obtained the services of Burfine was not precisely known either to Helix or the other members of the Board. The report was critical of the automatic control system and posed a series of difficult questions for which no ready answers were available. This report, a focal point in the controversy, illustrates the relationship that the Board had with its management.

Helix reports that his next step, after his discussion with the three engineers, was to go to Stokes with the Burfine report.

> I'm playing by the rules, so as a director I take the report to the General Manager without disclosing the names of any of the principals with whom I met.

> "Bill, I have this report," I said. "You know I've asked for that Board to be convened, reconvened, it wasn't, and I would really appreciate an answer to these charges. So I asked for this analysis, and I got called by Bill Stokes, asking if I could come to the BART headquarters to talk to people that were going to point out the weaknesses in the Burfine report.
>
> So, I said, "Sure!" I showed up at the BART headquarters and there were fourteen people sitting around a desk. Senior BART engineers, and engineers from PBTB. ...all the high powered types,...the heads of the divisions of the engineering firms. I was the only person there outside of the staff and PBTB engineers. They began to downgrade the report and said Beckers and Burfine were a couple of guys looking for a job. They're not very bright and they do a little engineering consulting.

After listening to several different engineers provide explanations and answers to the questions the Burfine report raised. Helix reports his reactions:

> I must have really been dense that day because I thought I was getting a big line of BS. I said, "You can pick out any concept in my field (education, originally) I can explain it to a child. Damnit, I'd have to believe you can do that with this system."
>
> And I said, "I appreciate all the effort and time that it took to compile this impressive grouping of materials and all the high powered individuals we have here, but I'm not satisfied. So I'm going to the Board." And mind you, I had not gone to the Board yet.
>
> I took it (Burfine Report) to George Silliman (the Board President) and George said, "Go after it, fine." I talked to members of the Board that I knew were in Stoke's corner. And they were saying, "Why do you rock the boat?" And they were telling me the same things that Stokes had told me. This is just a little outfit on the Peninsula. They don't know anything.

> And so I finally said, "You know I'm going to take this to the Board and if I can't get support I'm at least going to present it at a public meeting because damnit this is still a public railroad!...The people have a right to know that charges have been made that the system is not functioning properly, and that someday a train is going to run right off the end of the damn track."

Despite Helix's discussion with the Board members and his intentions of taking up the Burfine report with the Board before going public, it didn't work out that way. The charges contained in the Burfine Report first appeared in the _Contra Costa Times_, a newspaper that BART General Manager Stokes. was later to call a "garbage can newspaper, (viewed)...from somewhat of a professional standpoint." (Stokes, of course, had come to BART from a career as a newspaper reporter.) According to Helix, it was purely accidental and inadvertent that Burfine's report on BART appeared in the public press before he had a chance to present it for board consideration. Nonetheless, the appeal to the public left the impression among many that this represented an attempt by Helix to feather his political nest.

Helix recalls that a casual remark which he made to Justin Roberts, a reporter with the _Contra Costa Times_, was how it started. Helix had been critical of some of Roberts' reporting of Concord council actions. He goes on:

> And so at one of the Council meetings, when he was sitting down there, I told him so. And I said it in public. I felt that he had not accurately reported the Council's action and I thought he should reserve the editorial comments for that page. So Justin and I didn't start off on a very friendly footing.
>
> And then he was covering BART during some of those early meetings, and he said, "I have changed my mind about you. You're finally asking the kind of questions that someone should be asking as the Director." And he said, "The people in our county really don't know why BART

> is three or four years late in operating. They can't get any answers."
>
> After a Council meeting one night, I let the cat out of the bag and I did not do it intentionally, but I was talking about BART and I mentioned the Burfine Report. Justin Roberts said, "What Burfine Report?" And I said, "Oh, it's a report that I gave to the General Manager about the system." Well, he kept digging. I don't like secret meetings. I have felt in my public capacity, that, if you ask me the right question, I'll give you the answer.
>
> So he said, "Can I get a copy of that report?" And I said, "I'll tell you what, provided you make a commitment, I'll give you a copy of that report, but I'm going to ask you not to use it until I have a chance to get Stoke's response. He said, 'O.K."
>
> So he read it. I called him the next day, and he said, "This is dynamite. Can we use it?" And I said, "Justin, I'm going to hold you to your promise." So then I had that meeting I told you about, and it was a massive snow-job. I was upset. And he said, "Well, can I release it now?" And I said, "You might as well. This is going to be a massive cover-up, so you may as well let the people know what the hell is going on."

The series of Contra Costa Times articles appeared during January and February. The minutes of BART board meetings show that it was not until January 27, 1972, that the Burfine matter was assigned to the Engineering Committee, composed of Directors Rudy (Chairman), Anderson, Helix and Silva:

> President Silliman referred the matter of automatic train operators, including the computer (sic) and other hardware and the operating staff, to the Engineering Committee for investigation.

The Issue Goes to the Board

More than three weeks passed before the Engineering Committee met to hear testimony on the matter. By then, the issues had been very well publicized in the press. For Helix, the February 22, 1972 meeting of the Engineering Committee proved to be no more than a replay of his earlier meeting with Stokes and his staff. This time, however, Burfine, PBTB engineers, and the head of Westinghouse's project team, Dr. Woodrow Johnson and his staff, plus all the Board members, save Frank Alioto, were in attendance. Helix said of the outcome:

> I fell right on my fanny and poor Mr. Burfine, who was absolutely crushed and demoralized, and the engineers, who didn't want to show up at the meeting, were equally demoralized because they felt this was a chance to have some things corrected.

Helix, who was later "taken off the Committee," reports the February 22 proceedings this way:

> The day of the investigation of the Burfine Report came. In the the meantime I met with the engineers in the evening and I told them, I said, "Now tell me what you want to do. They're going to ask that you be identified." I said, "If you are identified, there's probably going to be an effort to terminate you. I'll do what I can, but I can't give you any assurances. So if you want me to maintain the confidentiality of who you are, I'll do it. I will not disclose who you are." They were worried, and with good reason as it turned out.
>
> So Burfine agreed to come to make a report. The meeting started at 8:00 in the morning, Thursday morning. The press were there and Burfine was there, but the engineers were not there.

> It began with an effort to try and get me to say who the engineers were. And I wouldn't. I might say it was a rather prolonged effort, with a lot of innuendos about that either they don't exist, or if they really believed in what they were saying or if they were really truthful they would be there to present the materials. And I said they are afraid they are going to be fired and that's where it is and I'm not going to disclose the names.
>
> Well anyway, Mr. Burfine was ripped to shreds, and so was poor Mr. Helix. We went from 8:00 in the morning to 2:00 in the afternoon non-stop. They even brought sandwiches right to the desk. All I can say is that I was horribly unprepared. I'm sure that my ignorance of engineering matters was obvious to all.
>
> I remember at the end of the meeting...we never got into the substance of what the hell the charges were. It was very slick. It was a lot of name calling and a lot of statements about trying to sabotage the system and long speeches about how horrible it was that a director would lend himself to such a shoddy procedure and all that. So everybody was--not everybody--some of them were dutifully outraged.
>
> I remember saying to a reporter when he asked me how I thought it went. I opened my coat and said, "Do you see the blood?" I thought I had been carved up pretty good.

The formal minutes of the Engineering Committee, however, have only this to say of the meeting:

> Subject to appear on Board Meeting agenda without comment. (February 24 meeting apparently) Mr. Asmus, PBTB Manager of Engineering, addressed the Committee reviewing the development of automatic train control. Dr. E. W. Johnson, Westinghouse V. P. and General Manager, Transportation Division, reported on the present status of train control work and plans for completion. Mr.Stokes presented the answers to 11 questions posed by Director Reedy and 15 questions posed by Director Helix.

> Following discussion with staff and Mr. E. Burfine, Beckers, Burfine and Associates, Director Helix moved that Mr. Burfine or a similar firm, be employed to pursue, with the General Manager and staff, the questions on automatic train operation for a report to the Board, with compensation by the District. The motion failed on a 2-2 vote. Director Silva's motion to have Mr. Burfine perform his services without compensation failed for lack of a second.
>
> Director Blake requested start date, scheduled completion date, and time extensions on the train control contract. Director Helix indicated he would contact unnamed BART employees to see if they will appear at Board meeting to present their specific concerns.

Two days after the meeting of the Engineering Committee the Board of Directors met to hear the report of the committee and to consider the affair further. Board minutes record the meeting as follows:

> Director Reedy, Chairman of the Engineering Committee, reported regarding a review of automatic train operation.
>
> Director Lange moved that the answers supplied by the staff be accepted, that a vote of confidence in the staff be expressed, and that this matter be considered closed. Director Anderson seconded the motion.
>
> Director Helix moved the adoption of a substitute motion that, consistent with the recommendation of President Silliman, the BART staff employees who feel that the problems have not been brought to the General Manager have an opportunity to bring these problems to the General Manager and Mr. Hammond at an early time. Director Silva seconded the motion, which failed. Ayes--5: Directors Bianco, Clark, Helix, Silva, and Silliman. Noes--7: Directors Alioto, Anderson, Blake, Chester, Doherty, Lange, and Reedy. Absent--0
>
> Director Clark moved that the original motion be amended by adding the words "at this time." Director Helix seconded the motion to amend,

> which failed. Ayes--4: Directors Bianco, Clark, Helix, and Silva. Noes--8: Directors Alioto, Anderson, Blake, Chester, Doherty, Lange, Reedy, and Silliman. Absent--0.
>
> The original motion then carried. Ayes--10: Directors Alioto, Anderson, Bianco, Chester, Clark, Doherty, Lange, Reedy, Silva, and Silliman Noes--2: Directors Blake and Helix. Absent--0.

And so, for Helix's effort to open up the whole issue of the automatic train control system, that was that. Helix notes:

> Unfortunately, the upshot of the meeting was a complete discrediting of the Burfine Report without ever getting into the substance of it. And I think a certain measure of discredit to me as a director fo having the temerity to bring this to the Board and not heed their advice to die at a quiet desk somewhere. So I felt badly.

Throughout the period of December, 1971 through February, 1972, then, Helix, with some little assistance of one or two other directors, had tried to bring the engineers' concerns to management. By the end of February, management felt it had dealt definitively with the issue, and so did the majority of the Board. During the two month period of January and February, the minutes for the Board show no other references to this incident other than those quoted for the dates shown above. The minutes, however, are replete with much other business, and it seems clear that this matter simply did not loom for the full Board as of great significance.

In reality, the Board's 10-2 vote of confidence in management, taken at its February 24 meeting, effectively closed this chapter of the incident for the Board, and for Helix, who recalls what happened right after the fruitless meetings he caused to happen.

> Anyway, a couple of nights after that, Bruder came to my house and we were sitting around and

> I was saying that it was too bad. ...I felt badly about the fact that now we had lost the opportunity to get into the Burfine Report because whenever the name was mentioned, Stokes, or anyone else could laugh at it and it would not be treated seriously. I said, "Damnit, I understand finance, I understand business, but I don't understand engineering. And I feel I've let you down." And he said, "Well, you ought to see what's happening with the contracts." So then I said, "What contracts?" He mentioned that there was a lot of hanky-panky with the contract.

This conversation, then, is the link between the Burfine--three engineers incident and another series of occurrences which, in the long run, had far more significant impact on the way BART conducted its business. On the basis of the conversation with Bruder, Helix was to go on to raise many questions about BART's contracting procedures, which did much to raise public and legislative doubts over the administrative behavior of BART's managers. These inquiries led to detailed investigations by the State Legislature which revealed many flaws in managerial practices and procedures. As a result of this legislative inquiry, substantial changes in the Board and in its role in policy making were called for and finally answered.

With the termination of Board involvement in the affair of the three engineers, management's efforts to identify the engineers intensified, and on March 2 and 3 the three men were fired. Helix does not know exactly how they were identified, but he believes that it was done through one document he had used that had the initial Hj on it (for Hjortsvang?), and by the selective release of information to the engineers that found its way to Helix.

A review of the Board's minutes for March reveals only one short reference to the firing. It was the last item considered on the March 9, 1972 agenda:

> President Silliman, at the request of Director Helix, referred the matter of over-all policies

> and procedures regarding the appeal process for employees and their dismissal to the Administrative Committee.

The Board had little to do with the firing of the engineers, and it is revealing that, even at this late date, Helix could get the support of only one other Board member, Bill Blake, for the position he took. The whole affair of the engineers, it appears, was a minor incident for the Board, which felt, correctly, that it was within the powers of the General Manager to fire personnel for cause. The real concern of the Board at this time was to settle the issue of a date for the inauguration of revenue service. The matter of the engineers was a minor eddy in this main current of Board preoccupation.

Chapter 9

Defending Professional Ethics:

Confused Role of Professional Societies

The ethical code of a profession is a statement of the ethical principles that can be applied to the day-to-day activities of its practitioners. In some cases, the code of ethics provides unambiguous guidance for the members of a profession. However, as with all codes that set forth ideal patterns of conduct, they often confront situations where the prescriptions for action are unclear or contradictory.

The Canons of Ethics for Engineers, as approved by The Engineers' Council for Professional Development, states that the engineer will have proper regard for the safety, health and welfare of the public. It also states, however, that the engineer will not disclose information concerning the business affairs or technical processes of an employer without his consent. It further states that the engineer will inform his employer of the adverse consequences to be expected if his engineering judgement is overruled.

Each of these principles is relevant for the particular case of whistle-blowing examined in this book. They will not, however, be interpreted in the same way or with the same emphasis by members of professional societies who are being asked to assist and support colleagues who are in conflict with their employers.

On Wednesday morning, February 23, 1972, as Bill Jones was dressing to go to work, his phone rang. When he answered the phone, Jones was met with a veritable barrage of words from a nearly hysterical Robert Bruder.

Bill Jones was President of the California Society of Professional Engineers. The only time he had met Robert Bruder was a little earlier that winter. In his role as President of the CSPE, Jones had attended a meeting of the Diablo Chapter, the chapter in which Bruder was an associate member. After the meeting ended, Jones had stopped into the bar for a drink before going home. He happened to sit next to Bruder and they had had a brief conversation. Beyond this casual chat, Jones neither knew nor knew of Bruder.

Now, on this February morning, Jones was shocked and surprised to hear from Bruder. He reports the content of his phone conversation:

> The gist of the situation was that he felt that his job (at BART) was in jeopardy. It would be a great distress to him to lose his job. He, as well as a couple of other fellows, names unspecified at this time, had been making noises that there were problems in the BART system and that it wouldn't work. Nobody backed them up; the superintendents didn't back them up. He was very, very upset indeed. He was--I'm try to get the right word--but one might say hysterical. It sounded to me on the telephone like he was almost in tears. He kind of wanted to lay the thing on my shoulders.
>
> And so I told him that I would contact responsible people from his chapter to get hold of him so they could sit down and talk about it.

Bruder, who was calling from a telephone booth on his way to work, was apparently so upset by the prospect of losing his job at BART, that it was very difficult for Jones to follow exactly what he was saying.

This telephone call which Robert Bruder placed to Bill Jones was not a request for any specific action by the CSPE. Rather, it was a general plea for attention and support. Jones' impression was that Bruder simply felt he was in trouble with his employer and that perhaps the President of CSPE could at least provide some comfort in this difficult situation. Bruder told Jones how he could be reached and Jones assured him that a response would be forthcoming and that someone would get in touch with him. In spite of the interest and the reassurance he offered, Jones was not particularly excited about his phone call; the thought passed through his mind that the fellow might have a personal emotional problem. He reasoned further that Bruder was a member of the Diablo Chapter and that, therefore, the Diablo Chapter people ought to look into it. Jones finished dressing and went to work.

When Jones got to the office he found a message from Bruder waiting for him. He had called the office at 8:10 and left a message saying it was urgent for Jones to call him, and that if Jones had not called by 9:30, Bruder would call again. Rather than call Bruder back immediately, Jones called Gil Verdugo, then President of the Diablo Chapter. Jones explained the situation and asked Verdugo to look into it.

Then, at 10:55 a.m. Jones called Jim Wright of the Western Council of Engineers (WCE). The WCE was then promoting and supporting organizations which represented professional engineers in their relations with employers--i.e. engineers' collective bargaining organizations. At that time, the WCE was actively attempting to become the bargaining agent for professional engineers working for a number of employers in the Bay Area. Although Jones wasn't sure if the WCE was working within BART, CSPE was at that time discussing the possibility of some kind of an affiliation between CSPE and the WCE. Jones called Jim Wright to find out if he knew anything about Bruder's situation, since Wright worked for the city of Oakland and, of course, the BART headquarters were in Oakland. Wright told Jones that he had, indeed, spoken with Robert Bruder, but that Bruder had not made clear just what the problem was. Wright further indicated that he was going to contact Bruder again and try to determine the real facts of the situation.

A little more than an hour later--at 12:10 p.m.--Gil Verdugo called Bill Jones and told him that, in his opinion, there did appear to be something worth looking into in this situation. After hearing this, Jones tried to call Bruder at 1:35 p.m., but he was told that Mr. Bruder was out for the day. Later that same day someone called Bill Jones' home and asked his wife to tell Jones that there would be a meeting at Holger Hjortsvang's house at Walnut Creek that night at 7:30 p.m.

Those in attendance at the meeting were Hjortsvang, Robert Bruder, Roy Anderson, Gil Verdugo and Jim Ripple (Secretary of the Diablo Chapter). Verdugo comments on the meeting: "We wanted to find out if Bruder was just a raving maniac or what? I didn't know him too well. He had called Bill and Bill had referred him to us. Bill said he couldn't understand what the guy was talking about." The consensus of the people at the meeting seemed to be that there was cause for concern, but Roy Anderson summarized their conclusions by saying, "We didn't quite see that there was anything that we could do at the time because he hadn't been fired, and there wasn't anything he was suggesting for us to do. And so we simply told him to keep us informed."

The same day that all this activity was taking place, the Bay Area newspapers carried reports of the meeting of the Engineering Committee of the BART Board of Directors, which had been held the previous day, Tuesday, the 22nd. These newspaper accounts reported the presentation of the Burfine report and its subsequent rejection by the Engineering committee. Among the interviews with some of the chief participants in this event was one with Director Helix. Helix was quoted as saying that the engineers who raised the issue "fully expect their employment to be terminated."

The next day Gil Verdugo attended the meeting of the BART Board of Directors, during which the report of the Engineering Committee was presented to the full Board. The report of that meeting, carried in The Oakland Tribune on Friday, February 25, contained the following paragraph: "Bay Area Rapid Transit Directors yesterday terminated their discussions critical of automatic train control equipment by giving their staff a vote of confidence in announcing the matter closed." Helix was quoted in this article as saying: "It's little more than a whitewash." But another BART Director, Arnold C. Anderson of Castro Valley took another point of view: "I have very little sympathy for the people (the three engineers) involved at this point. If I was running the organization, I think I'd fire them."

On February 28, Gil Verdugo wrote a letter to Bill Jones. Verdugo was obviously concerned, and did not feel that the issue was over and done with. In this letter Verdugo wrote, "We will continue to follow this situation with Bruder and hope to come up with some conclusions rather than close the subject as the BART-D Board did."

The next day, March 1, 1972, Bill Jones responded with a letter to Gil Verdugo. He wrote:

> I believe the important considerations in this matter are:
>
> 1. Having heard that some engineers think they have problems we (CSPE) are interested in resolving it;
>
> 2. That we take actions to resolve it if, indeed, it exists; and
>
> 3. That the engineers know of CSPE's interests and actions.
>
> I am delighted that you are keeping on top of the situation, and will give you, to the best of my ability, any assistance that might appear desirable or appropriate.

On March 2, 1972, Holger Hjortsvang and Max Blankenzee were dismissed from their BART jobs. On March 3, Robert Bruder was similarly dismissed. The anticipations of Bill Jones and Gil Verdugo were realized. The CSPE would, indeed, become deeply involved in the BART affair.

Part III

Aftermath

Chapter 10

Management Faces Public Scrutiny

Attempts to behave in accordance with written codes of professional ethics can frequently create serious problems and dilemmas for employed professionals. Real life situations rarely conform to the idealized situations contemplated in the codified rules of ethical behavior.

Ethical behavior is commonly associated with issues of conscience, that is, personal belief and commitment, rather than with provable and indisputable points of fact. Historically, individuals who insist upon behaving in ways acceptable to their consciences have had to be willing to pay the price of their convictions. Conscientious objectors to military service in time of war, for example, have been required to pay the price of alternate service or imprisonment, in order to conform to the standard demanded by their consciences. The question then arises as to what price--if any-- employed professionals can or should be required to pay for behavior which meets their ethical standards, but which violates some of the norms imposed by their employing organizations? This is an especially difficult problem when ethical abstracts are imposed in dealing with some of the ultimately unanswerable questions repeatedly posed in a highly technical society: How safe is safe in transportation or other mechanical technology? How pure is pure in food or water or air?

Indisputably, the public must be protected against careless, uninformed or venal behavior among decision-makers in organizations whose

> products and services the public use. How the ethical employee can help provide this protection is a question of increasing importance in our society.

On March 2 and 3, 1972, the BART organization terminated the services of Holger Hjortsvang, Max Blankenzee and Robert Bruder. As far as BART management was concerned, this was entirely a matter of internal personnel administration. Given the time it had devoted to the issue, management could not in reality affix the adjective "routine," to the incident, tempting as this might be. In management's view it should have been routine--just as the firing of any employee whose behavior is both dishonest and disruptive should be routine. But the actions of these three employees for the month or two before they were fired and for a long time after they were fired were perceived by BART's managers to have had so considerable an effect on employee morale within the BART organization and on the public perception of BART that it can hardly be called routine.

Actually, from management's point of view, it had taken a good deal of provocation on the part of the engineers before action was taken to dismiss them. Management avers that it had considered carefully the issues which were raised via the Burfine report and had repeatedly asked BART employees who had serious concerns to come forward to be heard. It was only after a continuation of the engineers' unprofessional behavior, compounded by their failure to acknowledge this behavior, that management felt constrained to take action. Yet, even this apparently temperate course has caused at least one actor in the proceedings, former Director Silliman, to accuse BART of being overly "military" in structure--undoubtedly because at least two of BART's high ranking engineers served in the U.S. Army Corps of Engineers prior to joining BART. Presumably, by "military" Silliman means authoritarian, rigid and unresponsive.

This is a particularly interesting charge to be made against BART management, especially in the light of testimony from the engineers themselves about the professional freedom which they and their colleagues enjoyed at BART. Repeatedly, engineers Hjortsvang, Blankenzee and Bruder speak of the lack of firm directives from management, and the freedom of employed professionals at BART to shape their own jobs. Not all of these descriptions of the working environment at BART are necessarily complimentary, but all of them suggest an atmosphere quite the antithesis of what is normally thought of as military in structure and tone. The question remains as to whether an organization can simultaneously be non-directive in substance but authoritarian in style. Finally, at least some managers of BART take the position, in the words of former Assistant General Manager David Hammond, that "there is nothing inherently wrong with the military structure of command and authority and responsibility if properly related to the type of employees that one might be provided with in a public agency management organization". (Personal Correspondence)

In any event, management, which was very much under the gun to inaugurate revenue service and was beset with the inevitable technical problems involved in coordinating a novel and complex system, perceived the firing of the engineers as no more than a difficult and unhappy employee relations incident, and continued about its work. It is for this reason that all of the subsequent discussion in the public press and in the journals of various professional societies went on without any comment by or contribution from management. Management claims that it did not wish the three engineers any ill, and its actions reflected this attitude. In the words of General Manager Stokes:

> (We refused) to talk about the situation because we considered it a personnel matter. The announcement, or news of the firing, came from

> the people involved and not from us. We did not comment in any way, shape or form. We did not (make) any efforts to--in any way--be surly to these three gentlemen. As a matter of fact, we hoped they would be able to go out and do their bit someplace else. We considered it strictly a personnel matter, and save for one time, I've never had any conversations with anyone about the circumstances of the whole firing incident. I wish it had not happened. It did happen. And I handled it the way I thought I had to handle it. And again, looking back even at this distance, I would, given the same set of circumstances, I would do it again. But all I want to make clear...is that we, at BART, despite what I consider to be some highly unprofessional comments by professional societies, operating strictly from one side, without any cooperations from us, because we just simply declined to be drawn into a discussion of it...(we) handled it just like any other personnel matter...
>
> We saw no purpose being served in terms divulging the specific reasons for firing, and this sort of thing. Again, on the basis that we hoped these gentlemen might go ahead and find other employment, (we) did no want to detract. We just didn't want them in the BART organization any longer.

This has been management's position since 1972. Others, of course, have involved themselves in the case for a variety of reasons. One example is Dr. Willard Wattenburg, one of several witnesses who gave testimony late in 1972 before the California Senate Public Utilities and Corporation Committee. Management's view of Dr. Wattenburg is that he is a self-appointed expert in automatic control, who was disappointed in his quest for a consultancy to BART. In General Manager Stokes' words:

> He had applied for--had suggested that BART, PBTB, whoever, should use his expertise, and it was a case of 'we'll call you, don't you call us' kind of thing. And I think he never quite recovered from that experience. As a result, (he) took on the role of gadfly...My God, every

> Board meeting, every time you picked up a paper, he was making some kind of charge.

Similarly, David Hammond recalls that:

> Wattenburg never really attempted to get in and find out what the design was supposed to accomplish or what the system was supposed to accomplish. Every time he talked he came up with a new approach, often contradictory to something he'd said before. I consider that he was much less than professional and certainly much less than useful in the solution of the problems that did exist.

As management sees it, then, one of the chief critics of ATC was apparently motivated by personal reasons to hurl charges and accusations at BART. He was, in their view, a disappointed office-seeker, and his self-esteem demanded that he castigate the organization which had refused to accept the offer of his services. Management's opinion of Dr. Wattenburg is not an isolated one. It is shared, specifically, by engineer Holger Hjortsvang, who wrote, in a July 21, 1973 letter to Professor Stephen Unger of Columbia University:

> Consequently, I feel slightly offended at being referred to as "one very honest engineer" in particular by a character like Dr. Wattenburg whose demogogic and irresponsible statements have done great harm by confusing the public, the Board of Directors, and officials with influence in the matter.

Like Edward Burfine, Wattenburg is characterized by Hjortsvang as a "jungle fighter", attempting to "gain fame by meddling in this sad affair."

Clearly, just as one's judgment of BART and its contractors or of the quality of their performance--however good or bad they may ultimately be proven to be--should not be based solely on the assertions of the Burfines and the Wattenburgs, neither should they be based only on management's characterization of the

Burfines and the Wattenburgs as men with petty personal axes to grind. Management, therefore, was gratified to learn that at least one of the engineers shared their view of the two consultants.

Although the role of the professional--technical societies in the months following the firing of the engineers does not have the high level of personal animus nor the quality of self-aggrandizement which may be seen in the behavior of some individual critics, still it does seem to management to be at least as much a manifestation of the attitudes of a few dominant individuals as it is a reasoned position of an entire professional group in support of an ethical principle.

Management's View of the CSPE

The first that management heard of the involvement of the California Society of Professional Engineers was When its President, William Jones, tried to call General Manager Stokes on March 13, 1978 concerning the firing. As it happened, Stokes was unable to talk to Jones, but Jones did discuss the matter with Assistant General Manager David Hammond. Hammond was frankly surprised that the CSPE had interested itself in what BART perceived as strictly a matter of internal personnel management. Because of this view, BART management did not agree to meet with representatives of the CSPE to discuss the issue. It was not that, on personal or professional grounds, such a meeting would have been particularly unpleasant for the BART management, for management was convinced that it had behaved properly and appropriately. Rather, it was a matter of principle that the precedent of meeting with outside and uninvolved groups on matters of organizational personnel policy and action should not be inaugurated.

In retrospect, Hammond adheres to this view. He has said:

> I have one piece of bitterness that I'd just as soon get on the record. That's the unprofessional conduct of the California Society of Professional Engineers. I don't believe professional organizations ought to act like unions; they ought to act like professional organizations...
>
> They were not at all interested in the problems or the solution of problems. They saw three engineers who were fired and they felt, like a union, that 'they can't fire our members'.

Asked whether there was a division of opinion in the leadership of the CSPE, Hammond went on:

> Very, very much so...We understood there were some other chapters (Golden Gate - ed) of the California Society of Professional Engineers who were very distressed at the action taken by this particular chapter (Diablo - ed). (Here Hammond is referring to the Diablo Chapter of the CSPE, which initiated a petition calling for the State Legislature to investigate BART and which otherwise provided support for and assistance to the three engineers).

General Manager Stokes shares this view of the proper role of the professional societies. He says:

> I don't see how the professional engineering societies can take umbrage at this fact (the firing) when engineers, put in responsible positions (and) living by the same code of ethics and code of operation that everyone else theoretically lives by--calling the shots in a responsible manner--can be criticized when people down the line have raised points and they're considered and rejected, how there can be any real problem. It seems to me that some of the professional societies are trying to have it both ways.

In short, BART's managers believed that the professional societies, to the degree that they undertook to champion the cause of the three dismissed engineers, inevitably cast aspersions on the professional competence and the professional integrity of the many engineers who were acting in supervisory capacities. If this point of view is accepted, it suggests that the behavior of the professional societies can itself be viewed as unethical, and, in fact, it was precisely this behavior which prompted the Golden Gate Chapter of the CSPE to criticize the Diablo Chapter and which led, in turn, to the passing of a motion of censure against the Diablo Chapter by the Executive Committee of the CSPE. This motion of censure was not, however, supported by an *ad hoc* committee of the CSPE which later concluded that Diablo Chapter did not violate the professional code of ethics by its activities in behalf of the three engineers.

The fact that charges of unethical behavior can cut both ways was recognized by some individuals who underwent a change of heart during the course of the whole affair. Edward Walker, for example, of the California Society of Professional Engineers, at first tended to support the position of the three engineers. But after talking with professionals on both sides, he changed his mind. He later reported his thoughts and feelings:

> It suddenly dawned on me, my God, we're (the CSPE) going out on a limb on one side of the fence and we're saying who were not truly registered and were not fully professional in *that* sense. Now they may have been honorable and so on...I assume they were honorable in all their activities and so forth. But the thing that concerned me was that, hell, there's a bunch of engineers, fully registered practicing professional engineers, members of the Golden Gate Chapter (of CSPE) who are members of the BART administrative and engineering staff. The people who let them (the three engineers) go were engineers. They were not administrators, strictly speaking. They were engineers. And I think it is wrong to go out and take one side or the other without finding out what the other side has to say...

> It's the kind of thing that you should weigh the facts and that is what I was hoping to do. In that regard...I think I was going to back off from an earlier position.

Eventually, after reviewing the entire case, Walker strongly supported a notion made by the CSPE to censure the Diablo Chapter of the Society for its one-sided actions in supporting the case of the three engineers.

Walker was not the only one. A number of other prominent members of the CSPE went through the same sequence of thoughts about the case. Even William Jones, who originally spoke to David Hammond in behalf of the engineers, later modified his position to the degree that he recommended to the Professional Engineers in Private Practice Group of the CSPE that they support the censure action taken by the Executive Committee of the CSPE against the Diablo Chapter.

It is not the place of management of a particular organization, of course, to get involved in the internal politics of the CSPE or, indeed, of any professional group--and its public silence on the matter reflects management's refusal to suggest what the motivations were for the behavior of particular individuals or chapters. But BART management was convinced that at least some of the activities of the professional societies in support of the three engineers were undertaken without a thorough or evenhanded exploration of the entire situation, and were more an emotional response to the perceived plight of three fellow professionals than a reasoned course of action taken after a balanced and searching review of the points-of-view of all participants in the episode.

One final word should concern the _amicus curiae_ brief filed by the Institute of Electrical and Electronics Engineers in the suit for damages which the three engineers brought against BART in 1974. Although management spokesmen have not commented directly on this matter, their total set of views on organizational responsibility and authority suggests that they would agree with the following analysis. The brief, which correctly limits itself to the issues of standards and codes of professional conduct for engineers and their relevance to the case at issue, cites the code's injunction to the engineer that he must "notify the proper authority of any observed conditions which endanger public safety and health." So far, so good. But the brief goes on to interpret the code to define the "ultimate proper authority" in the case of public employment such as BART as "the public itself." This is a sweeping conception which has the potential of undercutting the entire structure of corporate and organizational responsibility, even in public corporations, and which eliminates the process of orderly reference of problems to appropriate levels of authority and responsibility. Systematic delegation of responsibility is the only way in which such responsibility may be fixed, especially in the case of large and complex organizations. If properly delegated responsibility is to be ignored in favor of what is likely to be an emotional appeal to the public--and here the thorny issue is not pursued of just how the public as "ultimate proper authority" is to be reached--it becomes impossible to vest responsibility at the various appropriate levels of a public organization.

The Board Settles the Case Out of Court

The only additional role played by management in the wake of the firing of the three engineers was the legal one of giving depositions after the engineers had brought suit against BART for a total of $875,000 in damages. Various members of the BART management and engineering staff answered questions posed by attorneys for both sides in the case, offering their best recollection of the events as they had occurred. After a long period of this sort of preliminary activity, the case was settled out of court in January, 1975, with each engineer being awarded $25,000.

Even so, some highly placed members of management felt that BART should not have offered to settle even at this level. Erland Tillman, although explaining that he really did not know why the BART directors and their legal counsel chose to settle with the engineers rather than see the case to go trial, added that, on the basis of his knowledge of the facts of the case "unless they (the directors) knew lots that I didn't know, I wouldn't have (settled)."

David Hammond characterizes the settlement as "a capitulation on the part of the Board." He goes on:

> I know for a fact from having been called in to discuss the matter with BART's attorney that he was very confident that the outcome would have been in BART's favor, not in the three (engineers') favor. He (the attorney) was most distressed that he was under instructions to settle...I may be a little harsh in calling it a capitulation, because they probably saw it as something that they didn't need any more publicity on and if they could settle it and get it out of the way, they'd be better off. I wouldn't have seen it that way.

Although the fact of settlement may cast a slight shadow over BART's position, management believes, this must remain on the conscience of the Board, which decided, as Hammond views it, in favor of expediency rather than principle.

A Question of Ethics

The case of the three engineers has been widely discussed as an issue of professional ethics. As engineers themselves many of BART's top managers were and are deeply concerned with questions of professional ethical behavior, and are aware of and sensitive to the problems confronting the employed engineer who must deal with issues of professional obligations and responsibilities. The following analysis of these problems, while it does not reflect any points of view specifically articulated by members of BART's management, can be inferred from the position which these managers took on specific questions of corporate responsibility and decision-making in a large and complex organization.

More than many other professionals, engineers find themselves on the horns of ethical dilemmas not of their own making. Most physicians for example, are faced with ethical problems: shall they or shall they not tell the terminally ill patient the brutal truth? Shall they or shall they not continue heroic measures to sustain what is technically "life" in a patient? Shall they or shall they not recommend psychosurgery--with possible serious personality and behavioral consequences--for the hyperaggressive mental patient? These are all matters either of life and death or of issues almost equally grave, but they are issues in which the doctor is free--within the limits of his knowledge, his conscience and his patient's or the family's approval--to make and carry out his own decisions.

With engineers, it is generally different. With few exceptions, engineers are employed by large and complex industrial organizations. Within these organizations are various echelons of authority and decision-making power. Lines of authority and responsibility and channels of communication are necessarily complicated. As a consequence, the professional engineer is simply not his own master in the same way the physician is. The research division, the production division, the marketing division, the accounting office--all have some say over the character and direction of his work. Within the organization are countervailing forces, pushing this way and that in behalf of particular goals or interests, and the engineer may find himself squeezed between opposing demands of different groups in the organization. To crown the problem, most industrial organizations today view themselves primarily as profit-making entities. Although they may be perceived from without as automobile manufacturers or drug manufacturers or makers of industrial motors, if they do not show a profit for their stockholders, they will simply cease to be organizations at all. As a consequence, top management makes decisions--the impacts of which must be absorbed by the engineering staff--which are dictated by a multiplicity of organizational goals. The product design which the engineers may decide is best or safest or most durable may, for a variety of corporate reasons, not be the one that management chooses to adopt.

What is the ethical engineer to do? This is not a simple question to answer. At one extreme, he may simply insist to the bitter end that his solution to the design or production problem with which he is presented be adopted by the company as best or safest or most in the public interest. This sort of adamancy may work once or twice--especially if the engineer is a particularly proficient technical man or a master of persuasion. But it cannot always work. The demands of economics, of the market, of the company's production machinery may dictate that an alternative design or technique will be chosen. Convinced of the rightness of his position, the engineer may then say, "Do it my way or I quit." As a rule, the company will reply, "O.K. Quit."

At the other end of the spectrum, the engineer may lose sight entirely of his professional ethical obligations and respond with alacrity to every shift of the corporate winds. In doing this, he may indeed imperil public health and safety by not pointing out firmly the deficiencies of the particular product design which management chooses.

Obviously, neither extreme is a desirable course to follow. In the first instance, the engineer simply becomes so arrogant and convinced of his infallibility that he is unable to give fair consideration to the ideas of others and to grant the possibility that he might be mistaken. In the second, his concern with his own security within the organization subverts his allegiance to the professional standards which his status as engineer obliges him to uphold. BART management's position on the most appropriate course may be inferred from two policies it claims to have espoused with considerable consistency. First was the policy that all managers' doors were open to any and all employees at any time in order that comments and questions could be raised. Second was the belief that, in the final analysis, formal complaints had to be routed through accepted organization channels so as to preserve an orderly chain of command.

Different individuals obviously chart different courses between these two extremes of behavior, depending on their temperaments; their confidence in their own judgment, and the particular local context in which decision-making situations arise. It appears that the three BART engineers--to place the most charitable possible construction on their behavior during late 1971 and 1972--were so convinced of the rightness of their technical judgment as opposed to the accumulated opinion of some of their co-workers, their supervisors in BART, the consulting engineers of PBTB and the design people at Westinghouse, that they were unable to accept the fact that managerial opinion and corporate decisions did not always reflect their judgment.

Now this is not to say that they were wrong. Subsequent events have been somewhat ambiguous in elucidating the issue of right and wrong in the ATC System. Later consultants did, indeed, recommend a number of modifications in the system and the installation of a back-up control system, and these changes may certainly be interpreted as "proving" that the engineers were correct. Supporters of the engineers have pointed to two or three isolated incidents--a train overrunning a station, the accidental death of a track worker--as reinforcing the engineers' case. Still, the safety record of BART is outstanding. The problem was--and is--that the system is not as reliable as management would like it to be. It is the delays caused by redundant safety systems which tend to irritate BART users--not the physical risks of riding the BART trains.

Management has always granted that problems existed--as they exist whenever a new and complex mechanism goes into operation. Because the roots of some of the problems appeared to lie in design and fabrication of some of the rolling stock and some of the automated equipment, the BART legal staff and directors ultimately chose the standard way of registering dissatisfaction: BART sued Rohr, Westinghouse, IBM and PBTB. As is the general rule in corporate cases of this sort, BART's suits were answered by counter-suits brought by these suppliers of equipment and service. As of summer, 1977, the suits were settled, and the suppliers were ordered to pay BART certain sums in compensation for damages suffered. The settlement was similar to that accepted by the engineers in their suit against BART in that it amounted to only a few cents on the dollar asked. Management makes the point, however, is that there is little or nothing in these corporate suits which reflects or bears on the issues which the three engineers raised back in 1971 and 1972. Fundamentally, the operational problems which BART faced during its first five years of revenue service were problems of reliability. In the opinion of management, BART did not and does not have a problem of safety. But the system is designed so that, in the interests of safety, it tends to suffer a larger than acceptable number of delays in operation. BART's safety record over the past five years is undeniably excellent--and it is the issue of safety, after all, that the three engineers first raised in their objections, and it is to the issue of public health and safety that the IEEE _amicus curiae_ brief was directed.

Technical Decision Making

Basically, the view of management on the entire issue of ethical obligation and technical decision-making may best be represented by the following analysis:

Most complex decisions, in the final analysis, are judgments, and no individual's or group's power of making judgments is infallible. There always have been and undoubtedly always will be disagreements in judgment about technical matters. This is because technical problems can be solved in a variety of ways; rarely is there one way "right" to the exclusion of all others. And this is especially true in the development of any new system. As a matter of engineering philosophy, there simply is no single natural, pre-existing optimal design. A dozen competent engineers will come up with a dozen different organizations and arrangements of components--and it is only through constant feedback and a certain amount of empirical experience that modifications are made to improve the original conception. It was on the basis of this premise that management organized the BART engineering staff as it did--with the hope that it would get continual input. BART, like any novel electrical and mechanical system, had and continues to have problems, but the occurrence of problems diminishes in frequency and intensity as the system's engineers tackle them. Management knew--as engineers Hjortsvang, Blankenzee and Bruder knew--that the opening of the BART system would not be without problems. But the issue which management perceives as paramount lies in the character of the approach taken to these problems: it is not a question of correct or incorrect; rather, it is a matter of responsibility or irresponsibility.

Obviously the higher in the organization one gets, the greater the number of factors which must be considered in making a decision. If a particular technical solution to a product's safety problem makes the product so costly that it is placed out of reach of its potential users, then the problem can't really be said to have been solved. This is why management must look at the costs of solutions as well as their technical characteristics. Weighing and measuring trade-offs is part of the art of management--and this is precisely what BART management felt it was doing a lot of in the hectic months of late 1971 and early 1972. Admittedly, public safety is one thing that no manager should be willing to compromise--but opinions do--and presumably always will--differ as to when a product or system is "safe". And this is a point of view that the three engineers were apparently unable to accept.

This point is well made in an internal memorandum written by Mark Bowers, Director of Personnel at BART and a professional in the area of management behavior. He wrote:

> While District management is cognizant of the position of Hjortsvang and Blankenzee that they felt frustrated in their attempts to convince their immediate supervisor of alleged difficulties surrounding ATC, this does not lessen the severity of their actions. These employees, as staff engineers, are not classified or compensated to make management level decisions. Having presented their case to their immediate supervisor it then clearly became a management level decision to proceed with the information as management deemed appropriate. It was not for these engineers to assume without authority, for any reason, the role of management or attempt to usurp management's decision making responsibility.

Finally, some mention should be made of the 1972 and 1973 hearings of the Senate Public Utilities and Corporations Committee. In November, 1972, at the request of this Committee, Legislative analyst A. Alan Post issued a report containing criticisms of BART's ATC system as well as comments on a variety of aspects of BART's contracting and operating procedures.

Less than three weeks after this report was presented, BART issued a 157 page set of initial and supplemental responses to the report of the Legislative Analyst. In the opinion of management, Mr. Post's report was a mixed bag. It contained a number of valid criticisms and recommendations in which management fully concurred. But it also contained, management believes, petty quibbles, questionable criticisms and specific recommendations with which BART's management emphatically disagreed. In its reply to the Analyst's report, management tried to point out these recommendations.

Immediately after the report was filed, management went to work to try to implement those recommendations with which it agreed. The debate continued on those issues with which management disagreed. In this remedial effort, management worked closely with a special blue ribbon panel of experts, Drs. Bernard Oliver, Clarence Lovell and William Brobeck which was appointed by the Senate Committee.

In October of 1973, the Senate Committee reassembled to determine how well BART had progressed toward meeting the recommendations of the Legislative Analyst and a group of 21 technical recommendations which had been made by the special panel back on January 31, 1973. Alan Post reported that, although implementation of some of the recommendations had been started, he was extremely disappointed with the progress made thus far.

While management disagreed to some degree with the Posts' assessment of progress--as did its Westinghouse contractors--it did feel that Post and his staff operated with an acceptably professional attitude. The same can be said of Dr. Oliver, whose testimony is a model of thoughtfulness and moderation. In spite of the apparent efforts of Senator John Nejedly and some others on the committee to goad Dr. Oliver into intemperate criticisms of BART management, Dr. Oliver insisted that the problems were being adequately addressed.

> I am not conscious of any problems that I have had over the last six months or so in discussing in a factual manner the problems which BART is beset by. I am not conscious of the reluctance of the BART personnel to face these problems honestly and evaluate solutions. I think the response has shaped up very nicely in the last few months. I think they are honestly addressing these problems and they are receptive to worthwhile suggestions.

To which the Senator felt obliged to reply, "Why did they have to come so late, or do you have an opinion?"

Dr. Oliver, of course, refused to hazard an opinion, but simply said that he thought that the BART organization had been set up to receive a train control system "more functioning" than it actually got.

A little later, Dr. Oliver suggested that BART management had developed beyond its earlier optimism, and that, in facing train control problems, no longer had the attitude that "everything will be all right and it will be fixed tomorrow."

To which, Senator Nejedly, a representative of Contra Costa County, added, "Do you expect the BART management to make that statement in those words?"

Committee Chairman Alquist ended the interchange by saying that he did not want to ask Dr. Oliver to speak for BART management.

A little later during the hearing, Mr. Howard Miller, Manager of Engineering of the Westinghouse Transportation Division, asked to testify. Mr. Miller asserted that, in the opinion of Westinghouse, the equipment supplied to BART met "all our contractual requirements and all applicable industry standards for safety." Then, without letting Mr. Miller give his whole report, Senators Nejedly, Alquist and Behr subjected him to intense questioning which, in management's view, was clearly designed less to elicit information than to harass and embarrass. Even if the questions were motivated by a responsible and commendable concern for the public safety, the members of the committee were not above making political capital of this opportunity to be in the limelight. Even Senator Alquist was finally disturbed enough by the tone of the questioning that, at one point, he asked his colleagues to stop asking questions and permit Mr. Miller to give his testimony in orderly fashion, and added that it was not the committee's intent to embarrass Mr. Miller by asking him to comment on the report--Mr. Post's--which he had not been given enough time to study.

Finally, even though the behavior of some the senators during the hearing may be construed as self-serving, it is a model of good manners and decorum compared to the excesses of Dr. Willard Wattenburg's testimony, which followed immediately. Wattenburg exemplifies the group of independent critics of BART who took it upon themselves to act as representatives of the public in monitoring the development of the system. Unfortunately, in a number of cases, monitoring turned into a kind of perpetual harassing. Wattenburg, for example, began his testimony at this hearing by expressing the opinion that the Legislative Analysts' report and comments were "rather conservative." His own interest in BART was characterized by Wattenburg as "a hobby" which had "become sort of an obsession... better word." He went on to castigate the General Manager for "covering up everything and anything in sight," and asserted that "every colleague I know is snickering."

It had been hoped, during 1972, to improve the reliability of the communication and control system by equipping the cars with wheel scrubbers which would continually clean the tracks of the light film of corrosion which tended to cover them, and would make for better electrical contact between rails and wheels. The scrubbers, as it turned out, helped, but not as much as had been hoped. But of the introduction of scrubbers, Wattenburg could only say:

> The idea that you can clean the world to make it safe for your tender electronic trains is

> like saying you can keep Donner Summit free of ice with a hot water hose. That is the sense of the whole scheme. That is all there is to it. It is really taking a shovel with a rope handle, there is no evidence that shows those wheel scrubbers prove a damn thing.

Wattenburg was a man who, in management's opinion, was deeply disappointed because the BART system did not at any time hire him as a design consultant--even in the face of his self-promotion. He even went to the length of giving former Board President George Silliman "a document that I wrote that explained in layman's terms how this miraculous train control system works."

BART's managers suggest that it was difficult to maintain a professional level of discourse and debate in the face of public criticism of the sort Wattenburg continually offered.

To the degree that the PUC hearing was an earnest effort to discover from the legislative analyst and from expert consultants what some of the train control system's problems were, it was a legitimate and commendable activity. To the degree that it was a political affair designed to show off the rhetorical talents of some of the committee members and to serve as a public forum self-selected critics of BART, for it was not really designed to achieve its goals honestly or expeditiously.

In the final analysis, BART managers assert that they did respond to the criticisms and ideas of well-qualified specialists, and did make modifications in the train control system which provided adequate backup capability and which undoubtedly enhanced to some degree the safety features of the system.

But none of this is really central and relevant to the issues of professional ethics and organizational decision-making which the three engineers raised. The question which must finally be answered is, when has the professional engineer adequately satisfied the ethical demands which his profession imposes on him? Engineers Hjortsvang, Blankenzee and Bruder appear to have taken the position that their ethical imperative was not satisfied until everybody else in the organization had done what they recommended. The position of BART management--and, undoubtedly, the position of any large organization, public or private, is that the individual engineer has met his professional criterion for ethical behavior when he has as fully and as specifically as possible reported to his supervisor any questions, concerns, or reservations he might harbor about the products or the processes which the organization makes available for public use. If the engineer presents his ideas fairly--and if these ideas are given a fair hearing and due consideration by the decision-making echelon of the organization--no more can or should be expected of him.

Sometimes, the individual in a middle-level engineering position may be proven correct by subsequent events, and the whole hierarchy of decision-makers above him may be shown to have made an incorrect decision. But this, in management's view, is not the central issue. It is the correctness and the propriety of the process which is of ultimate importance if corporate responsibility and accountability are to be maintained.

Chapter 11

The Board of Directors:

A Changing Role

> Following periods of crisis, organizations will frequently make adjustments in the allocation of power and authority in order to adapt to changed internal or external conditions. Since crises often concern areas of responsibility or accountability, this shift is likely to be exemplified by an assertion by the nominal holder of power--typically the Board of Directors-- of its authority, as opposed to the <u>de facto</u> power which tends to gravitate toward managers who are in charge of day-to-day corporate decisions. In the case under study, the intervention of the public, in the person of the California State Legislature and its various committees, prompted the Board of Directors to take action it had earlier refused to take either because of timidity or lack of information. This move by the Board was supported by legislative action which made the change from an appointed to an elected Board. The question always persists, however, as to whether, in the long run, these changes will be substantive or merely cosmetic.

Engineers Hjortsvang, Blankenzee and Bruder were fired from their jobs in early March, 1972. Almost three years later, in January, 1975, the chain of their relationship with BART was finally broken with the out-of-court settlement of their lawsuit against the system. During these three year, circumstances, many of them directly and indirectly related to the events triggered by the actions of the engineers, forced some major changes in the character of BART's Board of Directors. Prior to the dismissal of the engineers, the Board appeared to be less in control of the organization's policies and directions than it was controlled by a variety of circumstances and by B. R. Stokes and his management team. After the firing, the Board swung--and was pushed--toward a more public oriented view of its responsibilities--one which demanded that it control rather than be controlled. The direct interest of the public in the actions of the Board was finally and formally recognized in October, 1974, when the Board became elective rather than appointive.

A variety of pressures--financial, safety, scheduling, legislative and contractual--applied from all sides, finally forced a change in the Board's role from an essentially passive one of accepting recommendations to a more active one of policy formulation for the BART system.

An examination of the highlights of this transition and of some of the major forces which stimulated it serves to show the firing of the engineers from the perspective of the Board, and to demonstrate that the pressures on the Board were as many and as intense as were those on the management of BART. From March, 1972 until the system's first trains ran in September, 1972, the Board was occupied with satisfying itself that the system was sufficiently safe and reliable to face the test of regular revenue service. During this same period it was also occupied with Mr. A. Allan Post, a legislative analyst in Sacramento. Post had been instructed by the Senate Public Utilities Committee to look into BART's financial status and operating procedures after Senator Nejedly, who represented Contra Costa County, pressed his Senate colleagues to support such an investigation. Once BART began operating in September, 1972, the Board's concern shifted to expanding service, removing the restrictions on operations imposed by economic reality and by the Public Utilities Commission, and obtaining the financing necessary to keep the BART trains running. Finally, the Board had to deal with improving its control of administrative practices, as urged by the State Legislature, with the resignation of General Manager Stokes, with suing its prime contractors for breach of contract, and with being sued by its fired engineers. All these demands on its time and effort motivated the Board to change its policy making role and the manner in which it exercised control over BART's management.

Post-Firing

Immediately following the dismissal of the three engineers the Board was remarkably quiet. Indeed there is no evidence the Board did any more than ask a few polite questions of its top management team about what had happened. Both before and after the actual firings, the general tone of Board comment supported management's action. Immediately after the February 24, 1972 Board meeting, for example, fully a week before the firing, Director Arnold Anderson (who was later to change his mind about such matters) stated publicly that the engineers responsible for the Burfine Report should be fired. Director Harry Lange later said of the firings:

> In general, it is my belief that this whole matter came up because these fellows were trying to get some recognition of their opinions, that they did it in the wrong kind of way that created both bad morale within the organization and a lot of resentment on the part of management and as a result of their perseverance along these lines they were fired...I believe it was their intention at the time to try to get some publicity that would have the effect of embarrassing management, which they did, and that was their objective.

Board President George Silliman recalled the firing this way,

> If I had been in Bill Stokes' position now, even though Bill and I were not seeing eye to eye, I think Bill did exactly what I would have done. I think from a manager's (viewpoint) you cannot condone, nor can you allow this kind of action within your command...I'm not saying these people were wrong in what they said, but from a management standpoint, Bill Stokes could not keep them in his employment.

When asked if, in his capacity as Board President, he had discussed the firing with Stokes, Silliman replied,

> No, I did not, and I purposely did not, because I thought this was a management operation and according to our law (the legislation that created the District) it was his job, his decision, and I didn't enter into it. I didn't even want to discuss it with him.

Insurgent Board member Dan Helix, who agreed that the firing was within the General Manager's prerogative was asked, "In the perspective of all the things that were happening was that (firing) a major thing?" Helix replied, "It was for me."

"But was it for the rest of the Directors?" Helix gave this response,

> Well, I think it served as an object lesson for other members of the staff. I think it effectively muzzled other members of the staff who were then starting to come forward in other areas. There was one about federal contracts where, along with the change order concept--if money was left over from one contract, a Federal contract for a particular purpose, it would be funneled into a different project, and I got into this...
>
> Question: After this incident...others at BART who had been coming forward...there wasn't as much evidence of that in your mind?
>
> Helix: I think it cooled off quite a bit. People were afraid to talk to me. I'd get anonymous phone calls you know.

With the passage of time, Helix's interest in the three engineers, like that of the remainder of the BART Board, dwindled. The Board, after all, did not have jurisdiction over BART's administrative and personnel matters. By law that power was left in the hands of the General Manager. Although this law was later changed to permit the Board to set administrative policies, including those which were to lead to better grievance procedures than existed at the time of the firings, in March, 1972, it was Bill Stokes' prerogative to fire the engineers. The Board's only alternative, if it wished to demonstrate its disapproval, would be to fire Stokes, and this they did not choose to do.

After the firing, the focus of Helix's interest swung from the engineers and the automatic train control system to the financial aspects of contracts awarded by BART, where his inquiries led to later embarrassment for Stokes and his management team. Through the use of his power to authorize changes in contracts, Stokes had been able to avoid Board involvement in fund expenditures of considerable size. In a hearing before the State of California's Senate Committee on Public Utilities and Corporations on April 23, 1974, Helix recalls his awareness of the fact that,

> On March 9, 1972, the Board was presented with a change order which had been routinely approved by the Engineering Committee. That change order was to have (sic) basic Concord-Richmond Yard contract. Items in the change order dealt with fare collection equipment and with escalators. We don't have escalators in our yards, and we don't have fare collection equipment in our yards. So when I asked why these items were in the change order, I was told they were there for the administrative convenience. I asked whether or not that was legal...About a month later we were advised that the change order was, in fact, illegal.

Justin Roberts, a reporter for the Contra Costa Times, became interested in these contract matters as he had been in the firing of the engineers. His year-long series of articles on BART in the Contra Costa Times was especially damaging to the credibility of Stokes and of BART's whole management team, and it undoubtedly helped lead to substantial changes in BART's management practices and in the role of the Board of Directors.

Revenue Service — When?

For Board President George Silliman, the firing incident led to a different set of concerns and raised a simple question: "Was the system safe? If these people (engineers) are right we're not going to run." If the system was not safe, Silliman reasoned, BART, which had already postponed its starting date several times, simply would not open. As recently as the Engineering Committee meeting which had heard the Burfine Report on February 22, 1972, the starting date had been projected by management to be in April. "If it would not open in April, when would it?", asked Silliman. He went on:

> So I decided I was going to find out...I grabbed Stokes, Hammond, Johnny Ray and said we're going to make an appointment with the Chairman of the Board of Westinghouse. We jumped on an airplane.
>
> We met with the operating vice-president, Spooner...We met with the President. I still wasn't satisfied. We met with the Chairman of the Board. And this was the most interesting thing you ever saw. I'm mad now. I said I want the truth...I'm not much of an entity and I'm causing all this trouble.
>
> Finally it boiled down to Burnham (Chairman of the Board of Westinghouse) asking Spooner if he could do it (open BART). Spooner said yes. Burnham said, "Well do it. Now what does it take to do it?" Spooner said, "I need X number of people."

> We set the opening date right there in those offices. September 7 or something like that. Well, we opened on the 11th.
>
> We had a plane load of technicians out here within a week and Burnham said to Spooner, Mr. Silliman and I are coming out and ride the test. And we did, he came out. I'll never know whether the test was real or not. But the thing ran perfect.

In spite of Burnham's commitment to have the ATC system functioning in time for a September 7 opening, local criticism did not die down. In June, 1972, Senator Nejedly, an increasingly critical observer of BART, proposed to the California Senate a resolution that a legislative analyst investigate BART's safety and financial circumstances. The resolution was passed and Mr. A. Alan Post, a tough, thorough, and highly respected staff analyst for the legislature, was assigned the task of investigating BART's financial and operating conditions.

Between the dates of Post's assignment in June and his report, October 9, 1973, BART did open. On September 11, 1972, revenue service began on a 28 mile section between Fremont and Oakland. Importantly, the opening of this single stretch posed no need to merge trains from the three other lines that were to comprise the BART system. And it was train merger which presented the really tough train control problems. In addition, to insure the safety of passengers on the single line being run, the Public Utilities Commission (PUC) insisted that BART use a "manual block system" to back up its questionable automatic train control system. This system required that observers phone ahead positively locating the trains from block to block on the line. This relatively primitive technique was hardly the innovation the Board had in mind in the sixties when it undertook to build a rapid transit system that would take advantage of the latest in space age technology.

Safety Issues - For Real

On October 2, 1972, less than a month after service began, a lead car ran off the end of the line into a sand embankment at Fremont, injuring five passengers, none seriously. But the accident the three engineers had been concerned about had now happened. While BART management minimized the gravity of the accident--it was laid to failure of a crystal oscillator, which Holger Hjortsvang later testified was an almost incredible occurrence--the accident provided critics of the ATC with the ammunition needed for further probing and renewed attacks.

One week after the accident, legislative analyst Post filed his report. It contained 31 major recommendations which stemmed from his criticism of the automatic train control system, the general safety level of BART and numerous administrative procedures of the organization. Post claimed that BART's entire structure did not properly consider the systems engineering problems it was facing, a point which seemed to reflect exactly the points which the fired engineers had been making.

Moving from safety to general administrative and financial matters, the Post report went on to suggest that BART had been overcharged by its contractors and that, in general, it would be unable to meet its opening dates for full systems operations without what came to be called a "fix" of the Westinghouse ATC system. In fact, the report urged that PUC not allow the remaining three lines to be merged until a much improved control system was installed.

BART management and its Board of Directors filed an extensive response to the Post report in which it concurred with many of its recommendations, but qualified most by stipulating that management rejected the implication that it was not already at least aware of and probably satisfying their intent. In short, management claimed, it had already anticipated the problems noted by Post, was not delinquent, and really had little it could learn from the October report of the analyst.

At the Senate committee hearing on the Post report, Board President George Silliman led off the BART response with this statement:

> We are very much concerned that the inferences in the analysts' Report have led to the wide conveyance of the impression that BART is unsafe. We, the BART Board of Directors, unanimously (sic) believe that BART is safe.
>
> We reaffirmed our confidence yesterday by adopting, by a ten to one vote, the following resolution:
>
> "It is the conclusion of this Board, after reading the Post Report and the answers that have been developed to it by the staff, that the District does have a safe system, that the system is operating safely, and that no extensions to the other lines will be made until the complete safety of each is assured; and, furthermore, that it is the intention of the District to maximize its efforts to maintain the highest degree of safety possible."

While the Board showed only near unanimity in its vote (10 to 1), rather than the unanimous vote Silliman claimed, later events revealed the mounting pressure on the Board. The year had begun with the engineers' questioning of the system's safety. It ended with the far broader and more credible questions raised by the legislative analyst and the California Senate Committee. Despite the Board's ten to one vote, there was growing evidence that even the Board was losing confidence in management, in their glib, but unfulfilled predictions of the expansion of revenue service and in their continuing but not well supported reassurance that all was well with the ATC. The complaints would just not go away. Dismissing the engineers had proved far easier than it would be to dismiss the concerns of the Public Utilities Committee and the California legislature.

Safety and reliability questions surrounding ATC and the transit cars themselves kept BART from opening the full 71 miles of track in 1973, including all important full service through the Transbay Tube, the link that was essential to achieving BART's full potential. Analyst Post estimated that BART was losing $1.5 million per month without the 220,000 daily passengers which Transbay service could produce. That expected level of service stood in sharp contrast to the 70,000 daily passengers who used the strictly East Bay service.

1973 Events

The Board continued to feel heavy pressures during 1973. First, there was the continuing plague of the ATC. Second, the transit cars themselves proved unreliable, with the result that maintenance became a source of delays and cost overruns, and an indication that BART management had simply not built an effective operating organization for itself. The continuing failure to find satisfactory solutions for the ATC and transit car problems led to further delays. With these delays came cost increases which were left uncovered by foregone revenues, while all the time inflation continued to eat away at BART's capital base.

The Board of Directors undertook several steps that showed its increasing impatience with its management group. It appointed a blue ribbon committee to recommend a solution to the track occupancy problem which was plaguing Westinghouse's "innovative" ATC design. Under protests from top management and Westinghouse that it was unnecessary duplication, work on a back-up system recommended by the committee was undertaken in 1973. Whatever its technical merits, the back-up system added costs and complications, and, in fact, did little to improve the quality of BART service.

The October 1972 Post report had been critical of cost control and administrative practices used at BART. The Board, therefore, contracted with the Arthur D. Little Company for consultation, and brought onto its staff Robert D. Profet, an advanced systems engineer on leave from McDonnell-Douglas Corporation, to serve as a technical analyst. Both the Little firm and Profet were to develop recommendations concerning BART's internal management and technical organization which were based on critical analysis of the system as they found it.

The major conclusion of the Profet report was that BART did not have qualified systems and electronics engineering talent at the management level. This conclusion echoed Post's earlier conclusions and recommendations, which, incidentally, had prompted Holger Hjortsvang to write to Dan Helix seeking his support in creating a new position for Hjortsvang as BART's systems engineering head. A systems engineering group was established in November, 1973, but not with Hjortsvang at its head. The other major outcome of the Profet report was that the Board which had brought Profet in decided to dismiss him.

The report of Arthur D. Little, issued in September, 1973, was also critical of BART's management structure and administrative practices and made recommendations for major reductions in planned personnel additions. By making comparisons to other transit systems, the Little report left no doubt that BART was being operated inefficiently.

Like 1972, 1973 saw criticism mount amidst continuing delays in Transbay service. Although the first train went through the tube in August, 1973, it would be 15 more months before revenue service to San Francisco would begin. These were trying times for the Board, which was being subjected to increasing pressure from Sacramento, from the public, and from self-appointed critics like Dr. Willard Wattenberg, an electrical engineering professor from Berkeley who called BART "Watergate on Wheels." At the same time, the Board was getting little satisfaction from PBTB, mount with no revenues or other funding sources to offset them.

1974 Events

The difficulties of 1973 seemed only to intensify as 1974 rolled around, and, as a result, it became clear that there was a growing estrangement between the Board and Bill Stokes and his management team. The outside studies and reports which the Board had commissioned in 1973 showed the Board's increasing interest in an independent assessment of BART's predicament. Votes that once were going 10-1 or 10-2 in favor of management recommendations were now just barely managing majority passage.

It became increasingly apparent, also, that BART's major problem was and was going to continue to be financial. Whatever the causes, and they were multiple, outside financial help would be needed if BART was ever going to operate as intended. The question wasn't so much whether or not help would come, but where it would come from and what strings would be attached to it.

In testimony given on March 19, 1974, before the California Senate Committee on Public Utilities and Corporations, Legislative Analyst Post stated:

> Now, what you are faced with today is that the District is currently projecting a financial deficit that approaches $100 million over the next five years. And this has been explained by the District management as largely attributable to rapid inflation, high labor costs that resulted from a labor contract, and that after all, this kind of a deficit is generally characteristic of public transportation.
>
> We would agree, from our analysis that higher than projected inflation and labor costs have contributed to the current dilemma of the District. But, on the other hand, we do believe rather firmly that in large measure the problems are attributable to controllable factors which were, and which continue to be, the responsibility of the District Board of Directors and District Management. And in our review of the District's administration during

the past two years, we have concluded that almost every aspect of the District's administration, from contract design, contract award, and contract supervision down to the daily management of operations and maintenance, have suffered from lack of direction and control, both on the part of the Board and on the part of management. And we think this lack of direction and control has not only resulted in an inability to provide the needed public transportation services but also brings into serious question the ability of the present Board and the management to provide adequate assurances of future solutions to their problems.

We believe that the District and State taxpayers can no longer afford the paralysis of management which continues to take place due to the management failures of the present general manager, coupled with the unwillingness of a majority of the members of the Board to assume responsibility for enforcing effective and proper management practices. I would point out that while a Board is not a manager and shouldn't act in the role of a manager, it simply cannot be so passive as to abdicate its responsibility to assure sound management. And we also point out that a replacement of the current general manager by no means is going to assure a solution to all of the District's present and future problems, but our view of the decisions that have made under his direction, showing his unwillingness to deal realistically with problems, and his actions in reducing the effectiveness of the Board have convinced us that the retention of the present general manager will assure that the present and future problems will not be resolved.

Post's testimony went on:

I want to say that no one can predict what the future's going to hold, but we do believe that placing an additional financial burden on the residents of the District, and perhaps, on the State, we believe that the Legislature has to have, simply has to have, strong assurance from the Board that the affairs of the District will be managed in a prudent manner rather than having them handled in what has been in the past too much of a public relations program.

Post's analysis included an indictment of the Board of Directors as well as Stokes.

> We think the Board has failed to ask tough questions of management and to hold management accountable. Instead, it's our feeling that the Board has abdicated its responsibility by allowing the District management to dictate the policy, and while management has been successful in dictating District policy to the Board, we feel the lack of top management, direction, and control has reached a point where it really pervades the ability of the entire organization to rectify its problems. We think that the tendency of the general manager to continually confront problems by diminishing their importance rather than approaching the problem by dealing with the real world has resulted in extreme frustration in many areas of middle management and, in some cases, this has really just turned into apathy.
>
> That leads us to the conclusion that if you are going to get out of this situation you simply have to have a fundamental change on the part of both the management and in the direction of the board itself.

Post's analysis and testimony had significant impact on the Senate Committee on Public Utilities and Corporations. At the opening of the fifth in a series of hearings on the economic and fiscal problems of BART, held on April 23, 1974, the Committee's Chairman, Senator Alfred Alquist, offered the Committee's conclusions based on Post's testimony in the previous committee session held in March:

> Before the Board (of Directors) presents its plan, I believe it is important, Mr. Bianco (President of the Board) and members of the Board, to make clear what the Legislature's position will be before any financial plan will be considered. Last week two statements of policy were circulated among all Bay Area Legislators which in essence called for new management at BART before any funding scheme would be considered.

> BART is now over four years behind scheduled opening and almost half a billion dollars over estimated costs. Would any reasonable person deny that at least a significant portion of the delay or cost increase is due to mismanagement? We have been reviewing BART's situation too long to contend that a management change is not necessary, and the task is now yours. Public funds will not be expended or authorized in an irresponsible manner. If you choose to close down your system in defense of your present management, the responsibility will be yours. The public can no longer tolerate either the excessive waste or the delay. To commit or authorize any funds without the assurance that past errors and policies will not be repeated would violate our responsibility, and I can assure that this will not take place.

It is significant that the Committee's position on BART management was stated by its Chairman before hearing the testimony of the Board of Directors, most of whom were in attendance for this session of the committee. The Board members were there to press for further financial support from the State in the form of the continuation of the one-half cent sales tax imposed earlier. The minimal price for the continuation, on the basis of Senator Alquist's statement, appeared to be a more prudent and involved posture on the part of Board and the dismissal of General Manager Stokes. Earlier, legislation had been passed that had altered the General Manager's "full power" over the administration of the District and enabled the Board to establish policies governing the operating affairs of BART. The Legislature had also acted to make the Board elective rather than appointive, with the first election to be held in the Fall of 1974.

Nello Bianco was the Board President in 1974. He expressed to the Committee the Board's feelings that neither suits against its poorly performing major contractors, nor the removal of Stokes was the answer, at least not the short run answer, to the financial dilemma facing BART. Critics, especially critics in the legislature, had been urging that the Board not delay further in pressing its damage claims for non-performance on the part of its major contractors. On the matter of the suits, Bianco stated:

> In my previous testimony before this Committee, I stated that the Board's policy was to continue to press for completion of major contracts without resorting to slow and costly court proceedings, but that we intended to constantly reexamine that position.

And on Stokes' removal:

> While I personally believe that a new general manager for BART is necessary, the near prospect of a system shutdown will not necessarily be changed by dismissal of the current general manager. Dismissal is a matter for Board decision based on overall evaluation of management performance. As you are aware, the Board is evenly split on this question.

The 6-6 votes on the question of Board confidence in Stokes and on other policy matters had sharply limited the Board's effectiveness, and even though Stokes' Board support was eroding, he still was able to have his way without clear majority support.

Bianco was at the hearings primarily to present the Board of Director's request for financial help. He went on to explain the Board's position.

> My main purpose in being here today is to respond to your request that the Board of Directors appear to present our plan for solution to the District's financial deficit...
>
> The Board of Directors has adopted the following motion which calls first for state support, and if necessary, if (sic) other combination of certain local sources requiring state enabling legislation:

> "The first priority for funds to sustain District operations comes from the state general fund, hopefully from tideland oil money. If that is not sufficient, the second priority be from bridge toll money. If that is not sufficient, third priority be given to motor vehicle in lieu taxes. If that is not sufficient fourth priority be given to sales tax on gasoline, and if that is not sufficient, fifth priority to be given to a temporary extension of the District's one-half cent transaction and use tax, with the understanding that the fifth priority be applied only after the first priorities have been shown to be insufficient in maintaining the District in operation."

As a practical matter, only the fifth priority ever had any real chance of acceptance. Senator Mills, a committee member who had already introduced a sales tax extension bill, discounted the first four priorities as politically unfeasible for a variety of reasons, and it was common knowledge that, unless the sales tax was extended, BART would simply close. The Board knew that. Everyone else knew that. Nevertheless, the motion that represented BART's alternatives for financing gave only fifth priority to sales tax extension.

Dana Murdock, a Board member, went to great lengths in his testimony to show that the cost overruns had five major causes, all of which were beyond the control of the current Board or its management: (1) substantial underestimates in the 1962 report for the bond issue; (2) higher than expected maintenance costs; (3) the $2 million plus annual cost of BART's police services where none had been proposed or planned for originally, (4) inflation, with the fare schedule remaining nearly unchanged, and (5) a labor settlement that increased costs $19.6 million over the three-year life of the contract. Murdock pointed out that inflation had been estimated at 3% over the life of the project, a rate that was only half the actual experienced rate.

Murdock was also critical of those on the Board and in the legislature who used BART for political purposes.

> It has been made very apparent to me since joining the BART Board in November (1973) that of all the special purpose districts created by the state, BART is perhaps the most political office, or at least a ready platform for those who desire to keep their name and image warm and visible to the public.

In a remark that raised a heated rejoinder, Murdock "shared" a quote that he found apropos to his short experience with BART.

> People who would carry over great public schemes must seek proof against the most fatiguing delays, the most mortifying disappointments, the most shocking insults, and worst of all, the presumptuous judgement of the ignorant upon their designs. (Edmund Burke)

Senator Nejedly questioned Murdock as to just who was ignorant, and later went on,

> Well, I would offer one comment. We are told over and over and over again that this system is the first of its kind and the first truly modern rapid transit system that has been attempted in this country in the last 50 years. We took the Committee to look at Toronto, Montreal, Boston, New York and some of the proposals in some of the other towns--Chicago, and I haven't seen any evidence to date that BART is that much more sophisticated and certainly not anywhere near as reliable as these others. And I can't say that you've got a great deal more to offer, so far, except a lot of talk and promises which are a long way from being fulfilled.

Of course, as usual, the ultimate power rested with those controlling the purse, and for BART that meant the California State Senate Committee on Public Utilities and Corporations. They held the Board and BART at ransom, and the price of the ransom was Bill Stokes' dismissal.

In May, 1974, Bill Stokes resigned as General Manager of BART. Most informed observers would say that the resignation was really a firing, and that the firing had been dictated, not by the BART Board, but by Sacramento. An interview with former Board George Silliman in 1977, three years after the event, illustrates the circumstances:

<u>Question</u>: You used the term "he was fired."

<u>Silliman</u>: Oh, of course. It was true.

<u>Question</u>: The pressure was there?

<u>Silliman</u>: Yes. The vote had been 7-4 or 8-5 or 7-5 rather.

<u>Question</u>: But he resigned. Was he given that choice?

<u>Silliman</u>: Well, I think that he decided to and he was in trouble in Sacramento. They weren't going to give any more money to this management.

Directors Lange, Chester, and others staunchly defended Stokes and pointed out the difficulties BART would have in replacing its General Manager at a time when the Board itself would very likely change dramatically as a result of the forthcoming election. This last ditch stand to save Stokes, however, proved fruitless.

The pressure that had mounted steadily over a dozen years finally cracked Bill Stokes and "his Board." Following his resignation came a rapid series of events that culminated in the choice by the electorate of the first elected BART Board and in the opening of the complete system in January, 1975. The Board filed its multimillion dollar suits for damages against its contractors in the Fall of 1974, and the election saw Nello Bianco become the only appointed member returned by election to the Board. The Stokes coalition had disappeared and the new Board was left with new problems.

Settlement with the Engineers

But what of the three engineers? Their suit for damages had been filed in the midst of all of this turmoil. It had been management's firm and consistent position that no settlement should be made with the engineers. But the events of 1974 and the enormous pressure placed on BART from all sides gave the Board a somewhat different perspective. From the Board's point of view the engineers' suit was no more than a minor skirmish, hardly worth the management and Board time it was consuming in Court depositions preparatory to trial of the case. Therefore, the Board, over management objections, decided on an out-of-court settlement, which was completed in January, 1975, shortly before the Transbay Tube was opened and full service initiated.

Former Director Harry Lange, a strong Stokes and management backer on the Board, when asked whether he believed the settlement as necessary, said,

> Oh, all we'd have been faced with is a lot of legal expense, a lot of time in court, a lot of time of the part of our management and there was no question that they did suffer some economic disability as a result of the thing (firing) and I felt that the way that it was finally resolved was the best management decision that could be made.

George Silliman also favored the settlement as he indicates in this exchange of questions and answers.

<u>Question</u>: Were you in favor of the Board's settling with these engineers out of court?

<u>Silliman</u>: Yes.

<u>Question</u>: Why?

<u>Silliman</u>: We would have lost the case.

Silliman believed that refusal to settle would have cost BART additional time and money on a case that might very well have been lost. Silliman contends fighting the case would have been foolish as there was little for the Board and BART to win, and much to lose, in diversion of time and energy badly needed elsewhere, as well as in dollars and public image.

From the Board's viewpoint the case of the three engineers never really amounted to very much. It was an incidental event and, however symptomatic it may have been of the basic flaws in BART's administrative structure, it never was worth much Board time and energy. Apparently the appointed Board wanted to wipe out these old problems by an out-of-court settlement and leave the new, elective Board with only BART's significant problems to confront.

The opening of the complete system many years after it was originally scheduled, at a cost well in excess of its original estimates, and in a context of genuine and widespread dissatisfaction, forced attention to just what it was about BART that permitted these things to happen. What was it that allowed Stokes to run his own show, often unchecked by the Board, by economic realities, or even by the State legislature? Was it simply the force of the man's personality? Was it the enabling legislation which sought to protect the General Manager from a politically motivated board that could potentially interfere with the operating decisions required? Was it the unnecessarily high degree of technical innovation built into the system? Was it the inappropriate interference of other local governing bodies in BART's affairs coupled with the system's lack of taxing powers. Or was it the combination of fortuitous circumstances alone which was responsible for BART's problems?

Whatever the causes, and they seem to have been many and complex in their interactions, BART was simply not well-structured for the enormous task it faced. The firing of the three engineers was only a footnote in the history of BART and yet it symbolizes and summarizes much that was wrong. Problems of safety, of timing, and of finance were abundant throughout BART's decade of development. That the engineers were technically right seems beyond question. But the inability of the Board and of management to respond positively and effectively to the engineers' concerns is symptomatic of the existence of a complex set of structural and administrative shortcomings which were built into the system.

All this leads to one root question: "What might have been done differently to avoid the delays, the cost overruns and the general administrative and policy difficulties which BART encountered?" The answer to that question is important, for if it could be provided, it would offer society knowledge about how it might best operate the increasing number of quasi-governmental businesses, the so-called "third sector" organizations, which combine characteristics of public and private enterprise to form a hybrid type of organization with a number of unique characteristics.

These properties can be illustrated by the following questions: (1) To whom should Boards of Directors of third sector organizations be responsible? (2) How can society judge the efficiency of organizations such as BART? (3) Can governmental bodies effectively plan long range undertakings in an elective environment that seldom takes a perspective longer than four years? (4) Can third sector organizations exist without fund-raising powers as BART was forced to do and which left it prey to intense political pressure? (5) Should technical, third sector organizations such as BART operate with lay people in policy-making administrative positions, whether at Board or high management levels? (6) Can third sector organizations be expected to be innovative in technical matters? (7) Can technical people be allowed sufficient independence from their base organization to permit open comment on issues of public concern such as safety and prudent expenditure of public funds? (8) Can multiple layers of government work together to permit design and implementation decisions to be made at the local level, while financial decisions are made at a more centralized level where fund raising through taxation has proven more efficient?

The engineers' safety questions were, in fact, never adequately treated. But the administrative structure of BART made that outcome likely, and the conduct of the engineers made it inevitable. BART management and the Board were as stymied by later events and by the pressure they created, much as they were by the safety issue raised by the engineers. The Board simply could not force management to react positively to the criticisms brought by an unsatisfied clientele. In the end, only economic pressure, which in a private business would be called bankruptcy, was able to force change. Finally, of course, it was the legislature which effected significant change by wielding the same power it had used to create BART in the first place. From the point of view of the Board of Directors, the circumstances of BART's creation left them, either by inexperience, by statutory distribution of power and responsibility, or by the nature of subsequent events and personalities, without the ability to establish firm policy and direction for BART. It would not be entirely true to say that, given BART's background and administrative structure, the outcome of the incident of the engineers was inevitable. But it would not be far from the truth. Whether the structural changes of 1974 would have made a difference had they been effected earlier is a tempting question, the answer to which will only come out of future BART-type organizations and their experiences.

Chapter 12

The Aftermath of the Firing for the Engineers:

Consequences and Reflections

> The Constitution of the United States guarantees to all its citizens certain rights that may not be violated by the government. Amendments to the Constitution provide for the rights of free speech, freedom of the press and religion, and due process of law as well as other rights that limit governmental power. But what about the rights of employees of private organizations? Should employees be able publicly to criticize their organization's practices or products without fear of retaliation? Should employees be able to write, and circulate to other employees and to appropriate publics, statements designed to get support for changes in the organization? Are the materials and papers in an employee's office or desk subject to the protection of rights of privacy or are they the property of the employing organization?
>
> Questions related to rights of employees and their employing organizations are often involved in incidents of whistle-blowing. Persons who oppose organizational authority and publicly air their criticisms frequently suffer either dismissal or other career-related deprivations. This delicate question of employee rights is a matter of increasing concern in both private and public organizations.

The reaction of the engineers to being fired was first shock and then anger. In their view, they had been led to believe that there would be no reprisals for the actions they had taken, which they believed to be simply the ethical behavior of professional engineers. Each of the engineers, in his own way, tried to make some sense out of what had just happened, and each may also have harbored the hope that the firing might be reversible.

On the evening of the day he was fired, Blankenzee called Daniel Helix to tell him of the day's events. Blankenzee told Helix: "I thought that, in every way possible, you were going to defend us and that we would not get fired if we would come forward." He also reminded Helix that he had related to them Mr. Stokes' statement at the meeting of the Engineering Committee (when Burfine got "slaughtered") that he would not fire any of the individuals who came forward with complaints about the BART system. According to Blankenzee, Helix thereupon told him: "Well, I'll tell you, I'll try to fight it as much as possible."

Two days later Helix called Blankenzee to introduce him to a friend, "a newspaper reporter who wants to talk to you and find out more about what happened on your firing." The reporter was Justin Roberts of the _Contra Costa Times_. Roberts, who was on an extension line in Helix's office, asked Blankenzee a number of questions about his being fired. This was the first of many conversations, both face to face and by phone, between Roberts and the three engineers concerning, at first, their dismissal from BART and, later on, the broader questions of BART's technical and management problems.

While the engineers seem to have talked to Roberts chiefly with the hope of regaining their jobs with BART, it is clear that both Roberts and Helix had their own special interests in mind. And these interests were basically unrelated to the employment of the engineers.

Once again, therefore, we see the engineers being drawn into an activity and being used for purposes which extended well beyond their own interests. This appears to be a continuation of a process that began when they first contacted Helix and thereby lost their personal control over the ensuing events.

A few days after he was fired, Hjortsvang called Mr. E. J. Ray to talk about the circumstances surrounding the termination of his employment. During the course of the call, Ray told Hjortsvang that Mr. Stokes and Mr. Hammond would be interested in talking to him.

On the morning of March 10, therefore, about one week after the firing, Hjortsvang met with Stokes, Hammond, and Ray. Although his recollections are hazy, Hjortsvang thinks that, at this meeting, he probably stated that he was interested in getting back his job with BART. In his own defense, he claimed that the actions of Burfine and Helix in going to the Board and in going public were beyond his control. In particular, he expressed the feeling that his influence over Helix had decreased sharply from the time that Burfine became involved as the outside expert. This attempt by Hjortsvang to regain his job came to naught.

The Reality of Unemployment and the Long Arm of BART

Although the formal relationship of the engineers to BART ended abruptly on March 2 and 3, the political controversy and turmoil which surrounded their dismissal kept them deeply involved for many months with BART and its problems. They continued to have contact with Helix and started a series of conversations with Justin Roberts of the Contra Costa Times, who involved himself in the BART incident as an area journalist. The social and political ramifications of their situation kept the engineers involved in the same issues and relationships that had occupied much of their time and energy in the proceeding months.

But after the first few weeks of playing in this larger arena, the engineers gradually perceived that being fired was, in addition, taking on a narrower, more personal meaning for them and their families. Indeed, it wasn't long before the loss of income and the inability to find new positions became the central experience of their lives.

Holger Hjortsvang

When Holger Hjortsvang was fired from BART on March 2, 1972, he was earning $19,500 per year. For six weeks he had no income. Then, about the middle of April, he obtained temporary employment as a consultant for a Los Angeles company--Daniel, Mann, Johnson and Mendenhall--which was involved in the design of a rapid transit system for the city of Honolulu. Hjortsvang worked for them for about ten weeks, during which time he commuted from the Bay Area to Los Angeles about once a week. His total income from this consulting job was about $5,000.

Hjortsvang had no other employment until May 14, 1973 when he began working for Moore Systems for an annual salary of $17,500 a year. Thus, for a period of about fourteen months Hjortsvang had a total income of $5,000, and when he finally did find work again, his starting salary was $2,000 below what he earned at BART.

Hjortsvang believes that his failure to find regular employment for fourteen months was not due simply to the unavailability of suitable positions. There is no doubt in his mind that it was the BART experience which was responsible for his 14-month long period of unemployment. For example, during the time he worked as a consultant for Daniel, Mann, Johnson and Mendenhall, he was offered a full-time position on the company's Baltimore office, where DMJM was the consulting engineer for the State of Maryland, in charge of designing a rapid transit system for the city of Baltimore. This offer was made by David Miller, Vice President of DMJM. But almost as soon as the offer was made, it was withdrawn. Why? At precisely this time, the Chief Engineer at

BART, David Hammond, had resigned his position to become a Vice President at DMJM assigned to the Baltimore office. It is Hjortsvang's opinion that Hammond simply would not agree to employing Hjortsvang to work under him at the firm's Baltimore office.

Max Blankenzee

Max Blankenzee, fired by BART on March 2, 1972, tried to find work with Disney Industries in Southern California, near the end of March. He did this by contacting Disney's software department manager, Dave Snyder, to whom Blankenzee was once assigned by BART to give Snyder a summary report on the BART system. Blankenzee had received a note of thanks from Snyder and his business card for future reference.

Snyder told Blankenzee that Disney was working on an advanced train system and that there might be employment there for him. Snyder set up interviews for Blankenzee with his superior, Jack Cornwell, and with the manager of the personnel department. Blankenzee told them of his experiences at BART and the reasons for his termination. He further provided the names of Ed Wargin and Charles Kramer as his immediate supervisors at BART, and included a letter of reference from Helix in his resume.

Snyder contacted Wargin at BART, and Blankenzee reports what Snyder told him about this talk.

> He (Snyder) stated that, as expected, Mr. Wargin had negative comments, but that he was expecting this to the extent, and that technical-wise Mr. Wargin didn't have anything to say, but that as far as political and management-wise, that more or less was naively involved in politics of the company and against management somewhat and he would disrecommend me again for employment.

Blankenzee also reported that he had heard from someone at BART that BART's Public Relations Department had been in contact with people at Disney, had labelled Blankenzee a "troublemaker" and had recommended that he should not be hired.

Following these exchanges of information, Snyder told Blankenzee that he could not hire him because Disney had decided to wait a few more months before starting the job for which Blankenzee had been interviewed. Blankenzee asked Snyder directly if BART had said anything about him that hurt his chances for the position, and he reports that Snyder replied, "we cannot answer that...we cannot take a position on this."

Blankenzee, continuing to follow the want ads in the newspaper, noticed in August that Disney listed openings for software programmers. He called Snyder again and was again told: "At the present time we have no need, do not need a person with your qualifications."

When Blankenzee was questioned about the fact that he had no direct knowledge of BART's contact with Disney and was relying on what someone told him, he replied:

> Well, I do not rely on what someone else just has told me. The thing is: A man knows when he is ready to get an offer and when he is not ready to get an offer. Disney gave all the implications of hiring me. In fact, we were talking about movers and which moving company to use and to get me down from here to the LA area; of the areas in LA where I could live, what the working conditions were and of my possibilities of having to cut my hair, shave my beard off if I have to go to Florida and work there for Disney. The indications were given by Disney that the job was mine...and all of a sudden, their attitude afterwards and the conversation with this gentleman (Snyder), I feel that BART blackballed me with that job.

Blankenzee believes that much the same thing happened to him in mid-May, 1972, he replied to a newspaper ad for a position with TRW Corporation. He sent his resume along with the letter of reference from Helix. Blankenzee was called to Portland, Oregon for an interview with a Mr. Wasie, who was in charge of the computer systems in the Bonneville Electrical Power Project. The interview was closed in the following manner, according to Blankenzee:

> (They said) "We would get in contact with you in the next couple of days. We want to check with BART and the other places that you have worked on your capabilities, for the record...Our personnel department will get in touch with you and probably make an offer if--after we talk to these people."

No job offer was forthcoming following this interview, although he admits that he has no information about TRW contacting BART, or vice versa, he feels once more that BART is responsible for his failure to get another job.

Finally, Blankenzee describes a third incident illustrating his difficulty in finding employment after being fired at BART.

In July of 1972, about four months after the firing, Blankenzee was called by a Mr. Sheldon of the Washington Metropolitan Transit Authority (WAMATA). According to Blankenzee, Sheldon got his name from someone at the California Society of Professional Engineers. Sheldon said they were looking for a person to do the central control computer systems for WAMATA, and "for a man who eventually could end up doing the managing of the computer center and the operations center."

Blankenzee received a letter from Sheldon describing the position and its salary range, and inviting him to Washington for an interview. Thereupon, Blankenzee flew to Washington and was interviewed by Sheldon, and by his superior Mr. Ralph Woods. Blankenzee describes how the interview was concluded:

> Mr. Sheldon stated at the end of the day that he had interviewed several people and that he would make the job offer to me after he had checked, because Mr. Ralph Woods felt that it has to be checked by BART or by Washington's

> chief engineer with BART's chief engineer that everything was all right.
>
> Mr. Sheldon made me the money offer, said about what the figure would be that the employment department from WAMATA would offer me, which was between seventeen-three and nineteen-four. And his conversation with me was: "I hope that I won't run into any problems with the management in convincing them to hire you."

Three days later Mr. Sheldon called Blankenzee, who reports the conversation as follows:

> Mr. Sheldon says, "I'm sorry Max," he says, "it was in my conversations with Mr. Kramer and the conversation...our chief engineer (had with) Mr. Hammond. They have advised us against hiring you."

Just before this incident, in June of 1972, almost four months after he had been fired, Blankenzee found work as a part-time consultant for Econolite. He held this position until September 11, 1972 when he took a job with General Railway Signal in Rochester, New York, at a salary of $16,200.

Blankenzee and his family had a difficult time adjusting to the move to Rochester. One day he saw an ad for an engineer with his qualifications at Philco Ford in Southern California, he answered the ad, and was contacted by phone and then flown out for an interview. As a result of the interview, he was offered a position at a salary of $17,500. He accepted the position on May 21, 1973.

Thus, fifteen months after he was fired by BART, Blankenzee was back in California in a position he liked and in an area that he and his family preferred. During this period, he experienced several months of unemployment or partial employment, a cross-country move that threatened the stability of his family, and unestimated but very substantial financial losses.

Robert Bruder

The impact of being fired by BART was greatest on the third engineer, Robert Bruder, both financially and personally. He describes how he felt and what he did in the first few months after leaving BART.

> There wasn't anything going on (in the job market). I don't really have much of a specialty. I'm not a design type...all the openings were...for specialists, design types, electronics, solid-state, or they were down in Santa Clara. I didn't even want to go down to Santa Clara. In fact, I didn't even want to go back to work. The heck with the world... I was drinking for a while there. I (worked for) a circus for a week. I bumped into sales jobs. I tried one of them. I lost more money than I made... I had no unemployment. I had some retirement from the funds out of the state retirement fund. I'd only been there (BART) a couple of years... I basically had no money from when I lost my job. People were hounding me for the house, this, that and the other thing. Well, we ended up on welfare in July. And it (was) sometime in July, I guess, or first part of August that I found this local job working for a small consulting firm... A deal which basically was paying the rent... Some relatives helped us with some of the monthly payments...so I put up my house for sale, you know, and got (it) off my back.

Specifically, Bruder had almost no income for five months after the firing, and for three more, earned only a minimal salary. The first job Bruder got after being fired on March 3, 1972 was with George Mathews Circus for about a week or ten days in late April or early May. He made about $20 a week doing what he described as "putting up the middle ring." He then began selling correspondence courses for LaSalle Extension, out of Chicago, a job that he started around the end of June and held until the middle of August. There was no salary associated with this job; it was strictly commission. Bruder reports that for these six weeks, his expenses exceeded his income.

On August 4, 1972, Bruder took a position with a consulting firm at a salary of $240 per week (approximately $12,500 per year). He held this position until November 13, 1972 when he was hired by Singer Business Machines at a salary of $1350 per month ($16,200 per year).

Since Bruder's salary at BART at the time that he was fired was about $20,000, the loss in income which he and his family experienced was substantial.

Bruder feels that the fact that he was fired by BART had much to do with his inability to find subsequent employment. Further, he feels bitter about promises which were made to him, and to Blankenzee and Hjortsvang as well, that they would be protected from retaliation or hardship if they "stood up" and put their case before the Directors. Bruder recalls some of the discussions with Helix at the time of the last showdown meeting with the Board of Directors. Helix told the three engineers that he also was under pressure to give up the struggle. As evidence of this pressure, he said he was offered jobs in San Francisco if he would resign his membership on the Board.

Bruder claims that it was at this time he realized that what he was involved in was more than a technical problem with BART's engineering system. In his words: "This was the real world...this is not Disneyland. This is it." However, Helix was reassuring about protecting the three engineers if they took a public stand. Bruder describes what Helix said to them. "(He) said 'Gentlemen, don't worry...It's still up to you. You will not be in trouble...You won't starve.'" Bruder interpreted such remarks to mean that he and his family would be protected. But obviously they were not.

Shortly after being fired by BART, Bruder went to see Helix to get his help in finding a job. Helix sent him over to the local electrical union where he was put in touch with "the top man at Bechtel personnel." This contact did not produce a job for Bruder, for reasons he explains as follows:

> I had an interview, and they needed people (in my) field. And I know I would have done the job. As soon as someone figured out who in the hell I was, I didn't even get a letter saying "No, we don't have any openings."

For Robert Bruder, as for his colleagues Blankenzee and Hjortsvang, it is easy to document the length of time he was unemployed or marginally employed. In his case, it was over four months before he obtained any stable employment as a consultant, and another four months before he obtained a regular position with a firm. But while the financial loss can be calculated, the effect of the experience on the man and his family is more difficult to estimate. Bruder feels that he was let down by people who promised to help in carrying on the "good fight." But he is still idealistic enough to believe that some good may have come out of the incident.

Contacts with Professional Engineering Societies

In the weeks just prior to the dismissal of the three engineers, Robert Bruder saw that there was a very good possibility of his being fired. Therefore, as a member of the Diablo Chapter of the California Society of Professional Engineers, he decided to try to get some help. One morning, he called Bill Jones, the state president of CSPE, whom he had once met at a meeting of the Society, told him of his situation at BART and asked for help because of his anticipation of being fired. Bruder didn't know if Jones would or could do anything, but reports that Jones called the Diablo Chapter and asked them to contact Bruder and find out what was going on at BART. Bruder believes that Jones called either Roy Anderson or Gil Verdugo.

Because it was impossible for officers of the Diablo Chapter to do anything while Bruder was still employed, he was not called by anyone from the Chapter until after he was fired, which was on a Thursday. The following Monday morning, representatives of the Diablo Chapter were at BART headquarters making inquiries about the firings.

Max Blankenzee was not a member of CSPE and reports no inclination to seek help from the professional societies after he was fired by BART. He indicates that it was Bruder who took the lead in contacting the CSPE, and who wanted the society to take a stand in support of professional engineering ethics. Blankenzee's initial anger against BART led him, rather than going to the professional society, to consider trying to find a good lawyer. But Bruder urged him to work through the CSPE.

About a week after the firing, the three engineers had a meeting with members of the Diablo Chapter and representatives of the Western Engineering Union. Bruder reports that Roy Anderson assembled the meeting at Gil Verdugo's business office, where "Roy had (called) his local representative committee and other local chapter people and other area people that were in the San Francisco area." The meeting was designed to uncover the facts of the firing so the CSPE chapters could decide on what position they should take. "We were called in one at a time. There were questions, answers, ...minutes, notes. They were trying to get to the bottom of just what in the heck was going on here."

Soon, one or two additional meetings were held between the three engineers and representatives of CSPE, especially members of the Diablo Chapter. Events moved quickly at this point. Helix became involved in these fact-finding meetings with CSPE representatives, and so did Justin Roberts, the journalist from the _Contra Costa Times_. Stories about the BART incident and the firing of the engineers appeared in newspapers and in publications of the CSPE. A state-wide fund raising drive to defend the engineers was led by the Diablo Chapter.

Bruder and Blankenzee feel that the Diablo Chapter was fully behind them, but that the Golden Gate Chapter, whose membership included members of BART and PBTB management, started to oppose the activities which were being launched in their support. Fund raising in behalf of the engineers was inhibited by actions of the Golden Gate Chapter, which initiated a move to censure the Diablo Chapter for engaging in unprofessional conduct.

This internal conflict within the CSPE continued for months. Blankenzee has reported that "we were told to stay more or less in the background and to let Roy (Anderson) and the leaders of the Diablo Chapter do the fighting." He recognized that, instead of a united front in CSPE, there were strongly divided loyalties between the management-oriented members and the working engineers. Blankenzee also perceived the development of a move toward compromise within CSPE, partly as a way of resolving the conflicts between the engineers and BART, but more important for the Society, between the chapters of CSPE within the profession. There appeared to be a growing consensus that a compromise was in order to avoid having the profession "air (its) dirty laundry in public." Once more, the engineers could perceive their interests taking second place to those of other actors in the drama.

In light of this move and this perception, the engineers decided to take their case to the courts.

The Law Suit

The decision by the three engineers to sue BART was based upon quite diverse motives. Blankenzee appears to have viewed a court case as the only way to bring out all the facts surrounding the actions of the engineers as ethical professionals who acted in the public interest, and the decision of BART management to fire them for these actions. In short, Blankenzee sought a public hearing in order to vindicate himself and his colleagues. Bruder and Hjortsvang, on the other hand, seem to have been more interested in recovering the financial losses which they had suffered because of being fired, and in exploring the off-chance that they might get their jobs back.

Bruder's recollection of how they made contact with a lawyer is that the suggestion came from the journalist, Justin Roberts. Blankenzee, on the other hand, has stated that Roy Anderson provided the contact with an attorney who was interested in taking their case. But even before this, Hjortsvang had already made contact with a lawyer on his own, hoping to find a way to get reinstated with BART.

Bruder and Blankenzee urged Hjortsvang to drop his course of independent action and join them in a common suit. Blankenzee recalls telling Hjortsvang: "We were in this thing together and the only way I thought we could win it is we stay together."

Preparations for the suit proceeded. Over many months, attorneys for the engineers and for BART took depositions from a number of the persons involved. But after all these months of extensive preparations, and with the court date only two months away, the engineers were called by their attorney and given some bad news. He reported on a recent court decision in which a public employee had brought suit because he had been fired. It appeared from the ruling that the judge decided against the discharged employee because he had lied to his superiors. The engineers' attorney feared that this case would serve as the basis for rejecting the engineers' suit, since they too had lied to superiors in denying any knowledge of or participation in the effort to reach the Board.

On the advice of their attorney, therefore, the three engineers decided to accept an out-of-court settlement of $25,000 each. Forty percent of each engineer's settlement went to the attorney, based upon their original agreement. Each of the engineers, therefore, won about $15,000 as a result of the recourse to the law. Bruder has rationalized his decision to accept a settlement by pointing to some of the things that had already been accomplished.

> Its almost two years after the fact (the firing) and this stuff is getting old you know. But really, who cares; and phone calls and people and publicity... as far as I'm concerned, let's settle for what BART wants and forget the whole thing. You know we've done our part...we know those things had been done (the friend of the court brief by I.E.E.E.), and of course there had been our other publicity in some of the trade magazines and one thing and another and this whole thing. I figured as a trade-off point we got to decide. You know. Let's wipe it out now.

Blankenzee was even more troubled by his decision to accept the settlement.

> To the end I kept saying no, no, no. We want this damn thing to come in court... I was getting tired of the runaround. Because I had so many depositions with the BART people. And then I decided look you know the whole thing is lost as far as the principle of the issues. The idea is not to get a settlement out of these people. The idea is to bring this case to court and make them see what kind of shenanigans were going on in BART. So that they would have to restructure things and that they should get some damn people in there that had a little bit of principle. But I guess we didn't, we couldn't, accomplish it on this basis.

Reflections

The BART experience had profound effects upon the lives of the three engineers. They suffered financial loss, altered careers, and disrupted personal and family relationships. They are even now, probably not fully aware of the scope and depth of the impact of that experience on them and on those they touched. What they will perhaps never be able to comprehend fully and to convey to others, are the fears and doubts experienced by individuals who take a stand in opposition to the impersonal and vast power of large organizations and institutions.

Reflecting upon their experiences, some six years after being fired, the engineers show some traces of bitterness. But they firmly believe that what they did was not only just and correct, but was, in fact, required of the ethical professional. Therefore, despite what they themselves experienced, they see some good coming out of the entire incident.

Bruder was asked if he thought a large part of the BART incident was a matter of misunderstanding, and poor communication.

> I don't call it poor communication. It's called the world we live in. Nobody wants to take (responsibility). How many people have the courage of their convictions? (He also recognizes all the reasons there are for not taking a stand.) You know, if I didn't have a ten-year pin coming next week I'd (get involved)...I would, but I've got to make the car payment or I've got to buy the house... And it all boils down to ethics as far as I'm concerned... That's the way I look at it. I says, holy smokes, you know you can't tell somebody else he should do this when something happens and stand up for your rights and what you believe, if you don't do it when your turn comes.

Blankenzee indicates that the only thing he would change, if he could do it over again, would be to try harder to get his concerns up the chain of command to top management. He feels that he should not have accepted the decisions of his immediate superiors in not permitting him to meet with top management, and should perhaps have written to them directly. As he put it: "That's the route I would have gone. If that would have failed, then I might have gone to Helix."

Aside from these reflections on the means used to contact top management, Blankenzee shares Bruder's view that the problem eventually comes down to the responsibility of the professional for the consequences of his work.

> What we are doing is something that has happened in wars, and happened in other things, and that says because this isn't really our job to do this, to bring this forward, we will close our eyes. That way we are not responsible. But I do believe that as engineers we are responsible. We are involved in the designs, we should be held responsible... for the designs in such a way that they should not inflict... harm on the public...Management's philosophy (is), "that's not your worry. You guys are just here to design..." But that's the difference between being a technical person and a professional person.

Bruder is very positive in his feelings about the role of the professional societies in trying to help the three engineers and in confronting the ethical issues involved in the BART incident. Blankenzee, however, has a very different view.

> The only thing that I can tell you about professional societies, and I have a bitter taste in it, the word professionalism, what they use, is a bunch of crap... Who are the people who are involved in the professional societies and the people that are on top of professional societies, are those managers who are causing the problems, and they want to put the lid on it. You know, that BART wanted to put the lid on us is one thing. But that the professional engineering society wanted to put the lid on us, and wanted to hunt us down is another... And so I really don't believe presently the professional societies are ready to fight for ethics. Those are words. Those are not actions, believe me. And you know, anybody from the professional society that comes up with all these rules (about professional ethics)...it's all right as long as they are in favor. But when they become in disfavor and they become quite controversial, well then I'm not going to back

> them up... That's mostly why I really am against the professional societies, and I think people like Roy Anderson, who think professional ethics are so high in his standards, I think he was kind of disappointed by the professional societies' actions.

Both Bruder and Blankenzee express sadness that the entire incident happened, but they say that they have grown stronger as a result. In Max Blankenzee's words: "I have to say it's sad that it happened. It happened, but we all learned from it, and I can say that I learned from it in the right way."

Chapter 13

Politics and the Professions:

The Limited Power of Professional Societies

> It is very difficult for a voluntary association of professionals--exemplified by the technical or professional engineering society--to sustain a firm course of action in the face of division of opinion among its membership, lack of resources, and inertial drag. In the BART case, the strong initial action of a chapter of the California Society of Professional Engineers stimulated a contrary reaction and eventually simply dwindled away, as time blunted the thrust of the activists in the organization. This inevitably raises the question of the ways in which a professional or technical society can play a useful continuing role in controversies of this sort. Both the purely technical and the abstractly ethical approaches seem inadequate. On the one hand, a technical advisory role does not permit the organization truly to come to grips with professional ethical matters. On the other, as this case demonstrates, the severe limitations of codified canons of ethical behavior make the society ill-prepared to deal with subtle and complex specific ethical issues.

Within two days after the firing by BART of engineers Hjortsvang, Blankenzee and Bruder on March 2 and 3, 1972, members of the Diablo Chapter of the California Society of Professional Engineers were beginning efforts to obtain redress for the three and to correct the technical problems they had identified. The CSPE engaged in intense activity on behalf of the engineers for most of the remainder of 1972, and continued their effort at a less intense level for more than two additional years. During this period, hundreds of phone calls were placed and scores of meetings held as the CSPE wrestled with the complex issues of the BART case.

In early 1973, the Institute for Electrical and Electronic Engineers became involved in this case, largely through the efforts of Professor Stephen Unger of Columbia University. Unger's interest in the case was picked up by the Institute's Committee on Social Implications of Technology, which drafted for IEEE's Directors a resolution of support for the three engineers. In response to the proposed resolution, the IEEE, through extensive committee work, finally filed an _amicus curiae_ brief in the lawsuit brought by the three engineers against BART. It is the activities of the CSPE and the IEEE during this period which this chapter chronicles.

The CSPE is Alerted

During the winter of 1971-72, Bill Jones, the President of the California Society of Professional Engineers, was involved in a series of discussions concerning the need for organizational representation of employed professional engineers beyond that possible within the existing structure of the CSPE. Most of these discussions were with Jim Wright, a member of CSPE who was also a member of the Western Council of Engineers (WCE). WCE was and is an organization attempting to promote collective bargaining for employed engineers.

Gil Verdugo's phone call to Jones on March 4 related precisely to the issues Jones had been discussing with Wright. Following this call, therefore, an emergency meeting was called for Sunday, March 12. On that date ten people gathered at the WCE Headquarters in Oakland. Attending the meeting were: the three fired engineers, Bill Jones (President--CSPE), Jim Lackey (President--Diablo Chapter CSPE), Gil Verdugo (State Director, Diablo Chapter),

Roy Anderson (Director of Diablo Chapter and Chairman of the CSPE Committee on Transportation), Ron Tsugita (President--East Bay Chapter), Jim Walsh (Director, NSPE), and Jim Wright (WCE). The purpose of the meeting was to present President Jones with all the information available at that time relevant to the dismissal of the engineers. Verdugo had told Jones that there was a need for some action to be taken at a level higher than that of the chapter; according to him it required state level support.

Bill Jones recalls his words to the assembled group. "We'll have this meeting, and you prove to me that there's something worth looking into, and we'll throw CSPE's weight behind the activity."

Following the meeting, Jones assembled his thoughts and began to act in his official capacity. He recalls:

> These three engineers had proper technical justification for their complaints within the BART System...There was public safety involved. Because of the actions they had taken, they had been summarily dismissed in a fashion which I thought professional engineers shouldn't be treated. The agreement at the time was that we must be prepared to go the whole way, to take it to courts to force some action. In the meantime, I would attempt to have the matter resolved by arranging a meeting with the General Manager of BART.
>
> On March 13, I called the General Manager's office quite a number of times. I never did get to speak with him. He was always out, or busy, or something, whenever I called. I eventually got a call back from Mr. Hammond as a response to my calls to Mr. Stokes' office. I told him what I, as President of CSPE, was calling about. In essence he said, "What the hell does this have to do with CSPE?" Those may be his exact words, but I won't swear to them.
>
> I could only respond by saying that the winds of change had blown, that the victimization of engineers was precisely what CSPE was concerned

> about, and that if the District Manager wouldn't talk with us we might have to pursue legal remedies.

Later in the week of March 13, Jones called Lew Riggs, the President of Tudor Engineering Company, one of three participants in the PBTB consortium which was, in effect, BART's general engineering contractor. Since Jones had known Riggs on a friendly basis for many years, he attempted to enlist his help in arranging a meeting between BART's top management and representatives of the CSPE.

Specifically, Jones wanted a meeting between Bill Stokes and himself. Riggs was receptive to the idea and volunteered his office in San Francisco as a meeting place for all who wished to sit down and to talk the issue through before it got out of hand. Sometime later, Riggs called Jones and reported that both Stokes and Hammond were totally unwilling to meet. Further, he informed Jones that the BART staff had been told that under no circumstances were they to talk about this particular affair with any member of CSPE. The order apparently came from the General Manager.

The imposition of this "gag rule" on BART employees prompted Roy Anderson to ask (in a telephone conversation with Bill Jones on March 26), "Can a public agency legally instruct employees not to talk about (public) affairs?" There seems to have been no clear answer to this rhetorical question.

Many meetings and phone calls followed the meeting at the WCE offices in Oakland on March 12. At the first of these, on March 14, nine CSPE members met with Justin Roberts (reporter for the <u>Contra Costa Times</u>) in Gil Verdugo's office in Lafayette, California. The CSPE members were: Verdugo, Anderson, Lackey, Tsugita, Walsh, George Humphrey (member, Diablo Chapter), Ed Walker (member, Golden Gate Chapter), Brian Tomlinson (Vice-President, CSPE), and Erwin Black (Engineering Supervisor with Cal-Tran and Roy Anderson's Supervisor in the Liaison Engineering Group overseeing the construction of the Trans-Bay Tube).

The members of CSPE felt some uneasiness about meeting with members of the press in general, and with Justin Roberts in particular. Roberts was at this meeting only because Max Blankenzee had invited him. Roy Anderson talks about Justin Roberts' presence at this meeting:

> We were in Gil's office; we were normal engineers, skeptical of newspaper people, never had any faith in them really. So we really didn't think Justin ought to be in there. We didn't know what he would report and of course this was the second go around with me (referring to his difficulties over press releases associated with his earlier stance on BART--See Chapter 5) and I had to keep a low publicity profile. I didn't want to get into trouble back at work again. Justin came in the front door, somebody said he was there, so I went out and kind of kept Justin out there for about an hour or so while the meeting went on. We began to talk a bit then and, more in the days and weeks that followed. I began to get some confidence in Justin, that I could have some trust in him, and I guess vice versa. We began to work together quite closely on this.

On March 16, Verdugo, Anderson, and Jim Ripple (Secretary of the Diablo Chapter) met with Hjortsvang and two days later Verdugo and Anderson met with Blankenzee. Verdugo reports on this meeting:

> He (Blankenzee) had a lot of wild stories. We had to sift through those, but there was a lot of truth in some of the things he said. He was quite emotional at that time too. He had been on a test train that went through one of the stations. The train didn't receive the slow-up, stop signal, and he said they went through the station with the doors flopping open at 60 miles per hour and scared the hell out of them. Things like that began to get us a little bit worried too.

On March 19, Verdugo and Anderson met to begin drafting a statement encompassing the information they had accumulated at their previous meetings. Additional meetings were held on March 20, 21 and 22, to develop the statement.

The Diablo Chapter Becomes Involved

On Thursday evening, March 23, the Diablo Chapter of CSPE met with Jim Wright, representing WCE, in attendance. The report prepared by Verdugo and Anderson was presented and discussed at length. The Chapter membership expressed their enthusiastic support for Verdugo and Anderson and encouraged them to continue their active involvement in seeking an equitable resolution of the apparent conflict.

At this point, Jim Walsh volunteered to attempt to set up a meeting between representatives of the CSPE and BART and PBTB management representatives. His offer was accepted by the members of the Diablo Chapter, but he faced the same reaction from BART management as Lew Riggs had, and was never able to arrange such a meeting. BART management steadfastly stuck to the position that the CSPE had no appropriate role to play in BART's internal personnel decisions.

On Saturday, March 25, Verdugo, Anderson, and Tsugita met again with the three engineers. At this meeting, they decided to have the three engineers ask BART to make a statement on why they had been fired. This request stemmed from an exchange of letters which had been initiated two weeks earlier when, on March 10, Max Blankenzee had written to E. J. Ray:

> I would like to appeal my termination with BART-D. At the present time, I am not sure if I will appear with counsel or not.
>
> Please inform me of the procedure I need to follow and my date for appeal. I hope to hear from you soon.

On March 24, Blankenzee received the following reply from B. R. Stokes:

> Your letter of March 10, 1972 to BART Director of Operations, E. J. Ray, has been forwarded to my attention.
>
> An independent investigation of your discharge has been completed by our Department of Labor Relations. Their report has been given my personal review as well as that of our Assistant General Managers for Operations and Administration.
>
> The result of this thorough investigation and review by top District Management indicates that your discharge for cause was warranted.
>
> Your appeal is therefore denied.

On March 26, 1972, Anderson and Verdugo met to draft another statement. This statement was an expression of their confidence in the testimony that the three engineers had given them to date. The purpose of the statement was to provide CSPE President Jones with the information that he might need in order to rally the support of the state organization on behalf of the three engineers. Jones received that report and responded by asking the CSPE attorney to take legal action on behalf of the three engineers against BART, and to examine whatever safety issues might be involved in the affair.

Elliott Pearl, the CSPE attorney, thereupon arranged for a May 10 meeting with the attorneys for BART, Pillsbury, Madison and Sutro. On that date, Verdugo, Anderson, Blankenzee, and one of the other engineers went with Elliott Pearl to the offices of the BART attorneys, but only Pearl went in to talk with the attorney for BART. After a time, he came out and indicated that, in spite of the fact the engineers didn't have a chance of winning anything, the BART lawyer had advanced the possibility of a 4 point settlement.

1. BART would re-hire the three engineers if they would agree to resign immediately. For this BART would -

2. Write them a letter of recommendation so that they would be able to obtain employment more easily and

3. Consider severance pay. (The engineers had received pay only for the last day they had worked.)

4. CSPE would appoint a three-man Technical Advisory Committee to assist BART in technical matters.

The BART attorney had stressed, however, that this was not a firm offer and that he would have to clear this with BART management.

But the tentative offer was not pursued and nothing came of it.

Roy Anderson comments on the events surrounding this offer.

> If we had had an attorney that had been interested, we might have been able to pursue that. If we had had an attorney who would have followed through immediately, within the week, and been tough, not condescending, maybe some pressure could have been brought to bear. I don't know, it was possible that Bill Stokes...may not have been agreeable to anything anyway. I think he thought he was in a pretty good position. He didn't think anybody cared. These were just three guys that nobody gave a damn about, and they had thrown people out on the street before and had had no bad results.
>
> We didn't think that it was too good a deal...but it was something that would have been up to them (the three engineers) to have decided.
>
> We didn't have a good attorney...that knew about labor relations, that knew about all these important things that you must do in order to protect your rights.

The Issue Moves to the State Level of CSPE

In May, 1972, the California Society of Professional Engineers held its annual meeting. Bill

Jones, as out-going president of the Society, used the full power of his position at this meeting to rally the state society to champion the cause of the three BART engineers. He gave an impassioned talk to the group, urging the full support of the statewide membership. Jones also gave the President's award to the Diablo Chapter in recognition of the work done by a number of persons in that chapter (particularly Verdugo and Anderson). The citation accompanying the award read, in part, "In recognition of its outstanding and extensive activities in springing to the defense of members of the Engineering Profession and the interest of the general public..." Finally, he urged the creation of a legal action fund to support the legal avenues to assist the three engineers. No immediate action was taken on this proposal.

Trying another tack in an attempt to relate the CSPE to BART and its problems, Bob Kuntz, the new CSPE president, appointed an _ad hoc_ committee in August, 1972, to confer with three BART representatives on general technical matters. This committee was to be under the general control of CSPE's Board of Directors and consisted of Bill Holden, William Tarman, and Billy Schmidt with Glen Wilson as chairman. According to Wilson, they were never able to arrange any meetings with management. The efforts of this committee represent yet another attempt on the part of CSPE to meet with BART and to discuss the situation in a reasonable and professional manner. The pattern in all of these attempts was the same; the CSPE people were willing to meet and talk, but BART management refused to do so.

Verdugo and Anderson carried on the bulk of the investigative work for the CSPE. In a short report, they summarize their investigative sources and their fundings as follows:

INVESTIGATION SOURCES

1. Retired BART officials, knowledgeable of management and fiscal matters.

2. Concerned test engineers working on the system.

3. BART's own audit reports.

4. Correspondence from other agencies pointing out areas of mismanagement.

5. Outside experts in the fields of computer science, train control, and fiscal management.

FINDINGS

1. Serious safety problems and unreliability of the system.

2. Engineering errors that have cost the taxpayers millions of dollars.

3. Mismanagement costing millions of dollars and years of delay.

4. Transfers and firing of audit personnel for reporting violations of state laws.

5. Intimidation of all employees by firing three engineers who spoke up in behalf of public health, safety, and welfare.

6. Withholding of information from the public on the serious nature of the automatic train control systems problems, its causes and effects.

7. Serious conflicts of interest.

CSPE Cooperates with Agencies of Government

Verdugo and Anderson describe how, within two months of the firing, their strategy began to shift from one wholly within CSPE to one involving various elements of government and the public.

> The material gathered was overwhelming, and the testimony heard made us aware of the seriousness of the matter. We asked for outside help

> and when we tried to explain the gravity of the situation, we were told that it was too big and too complex. The serious nature of the situation overcame the easy expectancy of "cooling it" as many of our contemporaries would have preferred. As we began to recognize the political overtones and the subtle comments made by our sources, our strategy began to change. We would see the cooling of some of the paths being followed, so we began to shift to the legislative trail.

California State Senator John A. Nejedly was the legislator selected by Verdugo and Anderson to be their champion. Verdugo had met Nejedly socially and their children went to the same parochial school. In addition, there had been occasions on which Roy Anderson had met with Nejedly and talked with him. Nejedly, a highly respected figure in Contra Costa County, had served as District Attorney prior to being elected State Senator. That service made a positive impression upon Verdugo and Anderson. They thought Nejedly had an investigative mind, and they believed that, when they explained the situation to him, he would be able to understand what they were taking about.

In the middle of April, therefore, Roy Anderson wrote a letter to Senator Nejedly. In it, he represented himself as the Chairman of the CSPE Committee on Transportation Safety. The letter pointed out the problems that Westinghouse was having with the reliability of its automatic train control system. Train detection and speed control problems were already apparent and concern was expressed that the three engineers had been fired because they tried to bring these problems to the attention of BART's management. The engineers had done what they had because they were concerned for the public safety, and because they were convinced that the system should not open until certain testing was completed.

In response to this letter, Senator Nejedly expressed interest and indicated that he had been approached by many people telling him that something was wrong with BART, but that these other people had never been able to substantiate their concerns.

A meeting was set for the Senator's office in Sacramento. The three engineers, Gil Verdugo, Roy Anderson, Bill Jones, and one other member from the Diablo Chapter (probably Jim Ripple) met with Senator Nejedly for about 45 minutes. The CSPE group made a careful presentation of their concerns and their preliminary findings. The Senator was convinced that there was substance to the claims and advised the CSPE to "get the facts together for a presentation to a group of legislators."

Throughout April and May and into June, Anderson and Verdugo worked diligently to accumulate the facts requested by Senator Nejedly. Anderson prepared a 34 page report entitled, "The BART-D Inquiry Preliminary Report." On Sunday, June 18th, Anderson, Justin Roberts, and Bill Jones met at Max Blankenzee's home to review the document that Anderson had written. However, before Roy Anderson could bring the report out of his briefcase for the others to examine, Bill Jones made it clear that he did not want the report to go to Senator Nejedly. For some reason, Jones was reassessing his position and began to move more cautiously that he had earlier. Anderson, therefore, left the report in his briefcase and the meeting became simply a general review of the events to date. After the meeting was over and everyone had left, Anderson took the report out of his briefcase. It was divided into two parts: Part 1 was 24 pages long and dealt with fiscal and contractual issues. The appendices included copies of audit reports, an excerpt on guidelines for engineering and management fees for civil engineering public works projects from an American Society of Civil Engineers publication, bid documents on the cathodic protection system for the TransBay Tube, a critique by Burfine and a critique by Follette. Part 2 dealt with train control and safety issues. Anderson asked Blankenzee to look through the report to see if the report was factually correct and to comment on how the information in the report might affect him.

When Roy Anderson got home that night he called Gil Verdugo and told him that, contrary to Bill Jones' wishes, he was going to forward the report to Senator Nejedly. He also suggested that Verdugo, who was now president of the Diablo Chapter, could create a Transportation Safety Committee for the Diablo Chapter and could appoint him (Anderson) chairman of the committee. Verdugo did just that, and the cover letter that was attached to the report was signed by Roy Anderson as Chairman of the Committee on Transportation Safety of the Diablo Chapter. The text of this cover letter is repeated below in its entirety. The letter conveys clearly the spirit of concern and professional responsibility that was exhibited in all of the actions of the personnel of the California Society of Professional Engineers.

Senator Nejedly used Anderson's report as an information base for discussions with other state legislators, but the response of his fellow legislators at this stage was relatively cool.

A few weeks later, around July 10, Justin Roberts, Gil Verdugo, and Roy Anderson met with Senator Nejedly in his Walnut Creek office. Anderson quotes the Senator as saying: "I don't think I'm getting any support. I don't think we're going to be able to make it." Verdugo's recollection of the Senator's comments at this meeting are somewhat more explicit: "Hell, nobody's going to believe you, not until we have a transit rider skewered on a rail, can we do anything. We can't, nothing's happened." As the group continued their discussion, a change in tactics appeared to be necessary. Verdugo's comments again describe their feelings at the time. "Nobody understands safety, but they sure as heck understand money!" Anderson recalls saying, "Well Senator, what if we had public support? What if we could start a petition and bring in names to

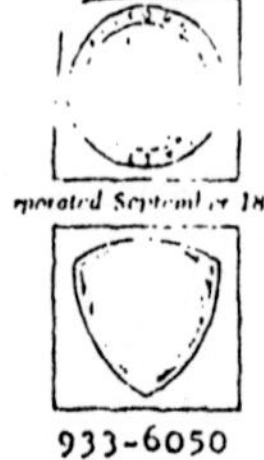

ncorporated September 18, 1947 California Society of Professional Engineers
DIABLO CHAPTER 18

3190 Old Tunnel Road
Lafayette, California 94549

933-6050

June 19, 1972

Senator John A. Nejedly
1393 Civic Drive
Walnut Creek, California

Dear Senator Nejedly,

This preliminary report "THE BARTD INQUIRY" is submitted to you for your use in bringing about what we believe is a most necessary and long overdue legislative investigation of the management and conduct of the Bay Area Rapid Transit District.

The contents of this preliminary report are based on testimony given to the Committee on Transportation Safety and from documents that have been gathered from many sources. Bart's instructions that their employees should not talk with CSPE has, of course, made our inquiry exceedingly difficult. The intimidation of all BART employees by Mr. B. R. Stokes and the BART Board of Directors, through the firing of the three engineers who brought their concerns for safety to the Board, is at best a suppression of a professional's responsibility.

The following is a summary of the more important aspects of our inquiry:

1. Unnecessary redesign of Transbay Tube seismic joint and payment for faulty material, cost an unnecessary $3.6 million.

2. BART's consulting engineers, PBTB, employed persons as engineers that were in violation of the PBTB contract with BART. Charges were made for services not rendered.

3. Purchasing procedures by BART staff in violation of the law.

4. An apparent conflict of interest by law firm which represented both BART and PT&T, gave a $6 million windfall to PT&T.

5. The estimated engineering fees of $56 million have now cost the taxpayers $137 million.

6. Corrosion protection equipment to protect Transbay Tube from destructive corrosion has never worked correctly.

7. Design errors have cost the taxpayers millions of dollars and have greatly delayed the beginning of system operations.

8. The safety and reliability of the transit system is in question. Lack of systems engineering and poor quality control have caused excessive delays, excessive cost and a system many years overdue. No date for automatic safe and reliable service can be accurately predicted while the system remains under the control of the present management. Major technical problems still exist.

We do not submit this report with great pleasure. We feel that this transit project is necessary and desirable, however, the findings of our inquiry indicate that major instances of mismanagement and misconduct on the part of BART staff and the consulting engineers (Parsons-Brinkerhoff-Tudor and Bechtel) are in evidence. We believe a thorough investigation by the legislature will reveal that proper management, both technically and financially, would have produced an operating system many years prior, at a much lower deficit than has been incurred to date. The facts do not reflect favorably on either elected officials of the three county areas or on the engineering profession as a whole. It is time that we correct our mistakes.

The transit system is still not operating. This means further unnecessary cost to the taxpayer and of course, even worse, no service. Our investigation shows serious problems in quality control that will no doubt affect the safety and reliability of the system. We wrote Mr. Silliman, BARTD Board President, on June 5, 1972 suggesting that they consider starting service under manual control in light of their many unresolved problems. We also suggested that an independent consultant be retained to review the system as designed and installed to assure the public of the system's safety and reliability. His reply on June 13 indicates that all is well and he appreciates our interest.

In closing, our committee would like to express our thanks to those three engineers, Mr. Blankenzee, Mr. Bruder and Mr. Hjortsvang who so unselfishly sacrificed their jobs and personal well being on behalf of the public health, safety and welfare. We believe the citizens of the Bay Area also owe them a debt of gratitude.

We urge you to do all possible to bring about an immediate legislative investigation of all areas of BART management and conduct. We would further like to thank you for your past help and interest.*

Sincerely,

Roy W. Anderson, P.E.
Chairman, Transportation Safety

RWA:jm

*Reconstruction of final paragraph above: "We urge you to do all possible to bring about an immediate legislative investigation of all areas of BART management and conduct. We would further like to thank you for your. . . help and interest."

you? Do you think that would help?" The Senator replied, "Yes, I think that would help."

Within two days, Gil Verdugo and Roy Anderson had drafted a petition for public support, and the Sunday, July 16, 1972 Contra Cos-

---

The California State Legislature is requested to conduct a

BARTD INVESTIGATION

to determine the facts on the following:

1. Why the taxpayers must pay $1.4 billion for a system that was to cost $792 million?
2. Why the SYSTEM that was promised to open in 1967 is still NOT READY?
3. Why the Consulting Engineers FEES have soared from their own original estimate of $56 million to more than $137 million?
4. Why the TAXPAYER PAYS for all the ERRORS of the Consulting Engineers?
5. Why the GENERAL MANAGER is to be paid a salary of $68,000 to direct a system he cannot get into operation? (This is more than the salary of the Governor.)
6. Will the SYSTEM ever WORK Safely and Reliably?
7. Why BART is NOT ACCOUNTABLE to any State or Federal agency or to the taxpayers?

The undersigned CITIZENS and TAXPAYERS of ALAMEDA, CONTRA COSTA and SAN FRANCISCO counties REQUEST the Legislature to CONDUCT an immediate INVESTIGATION, including PUBLIC HEARINGS and prepare a written report for the public review.

The Legislature is also REQUESTED to APPROVE the Bill by Assemblyman Carlos Bee to REQUIRE the BARTD BOARD OF DIRECTORS to be elected by the BART District voters.

CITIZENS OR TAXPAYERS

	NAME	ADDRESS	DATE
1.			
2.			
3.			
4.			
5.			
6.			

INSTRUCTIONS:

Signer of Petition DOES NOT have to be a registered voter. Any resident, 18 years and older, of Alameda, Contra Costa or San Francisco counties may sign. Return no later than September 30, 1972.

Return petition to:
Diablo Chapter, California Society of Professional Engineers (CSPE)
3190 Old Tunnel Road, Lafayette, California 94549

(415) 933-6050

ta Times featured a front page article by Justin Roberts on the petition drive. The strategy was now to go public; to generate sufficient public support for action that the state legislators would be more willing to support a legislatively sponsored investigation of the BART activities.

The public response was fantastic. Verdugo was inundated with requests for copies of the petition. Boxes full of signed petitions were collected and compliments for Anderson's and Verdugo's efforts poured in. Anderson and Verdugo were invited to speak at civic and service organization groups through the county.

The strategy had worked. Only five days after the public announcement of the petition campaign, 17 San Francisco Bay Area legislators signed a letter authorizing A. Alan Post, the Legislative Analyst, to conduct an investigation of the operations of the Bay Area Rapid Transit District, with particular reference to safety and contract administration. The appeal for public support had been dramatically successful.

The Legislative Analyst was charged with determining the following:

1. What problems affecting the safety of passengers still remain unsolved, and in what manner can the legislature assure that proper steps will be taken to resolve these problems?

2. What were the deficiencies, if any, in administration or the governance of BART which should be reported to its governing board by the legislature for corrective action?

3. What recoveries should the Attorney General attempt to obtain from contractors, engineers, or consultants?

4. What additional steps should be taken to assure that effective reforms are carried out?

The report prepared by the Legislative Analyst is dated November 9, 1972. On its opening page, the following paragraph appears.

> While we based our initial investigation on the study submitted by the Walnut Creek (SIC) Chapter of the California Society of Professional Engineers (Dated June 19, 1972) we have expanded its cope to assume a valid basis for definitive conclusions.

The 106-page report contained 31 specific recommendations. The largest number of these--10--had to do with the examination contracts. Eight recommendations dealt with passenger safety considerations and seven dealt specifically with the automatic train control system.

Five days later, on November 14, 1972, the Senate Committee on Public Utilities and Corporations held the first of at least 14 public hearings specifically devoted to Bay Area Rapid Transit District. This Committee was chaired by State Senator Alfred E. Alquist. The hearing on November 14 was devoted to a detailed presentation of the substance of the Post report and a discussion of the effectiveness of the Public Utilities Commission in maintaining appropriate safety standards. (Members of CSPE had met with the California PUC on June 8, 1972 in San Francisco to discuss the safety of the BART system).

The following week, on November 21, 1972, representatives of BART were given an opportunity to respond to the Post report. George Silliman, then President of the BART Board of Directors, began the testimony on behalf of BART. Silliman was accompanied by a majority of the Board of Directors and by several persons from the management staff. Discussions with the BART staff took up an entire morning session. The afternoon session saw Mr. Post and Dr. Willard Wattenburg testifying before the committee.

This second hearing concerned itself with the first 15 recommendations of the first Post report and resulted in an authorization by the

Rules Committee of a three-member panel of experts that would evaluate BART's automatic train control system. This Blue Ribbon Committee was composed of B. M. Oliver, Vice President for Research and Development, Hewlett-Packard Co., W. M. Brobeck, President of William M. Brobeck Associates, and C. A. Lovell, an Engineering Consultant. The committee was charged to report back to the Senate Committee on Public Utilities and Corporations not later than January 31, 1973.

The third legislative hearing of the Senate Committee was held on November 28, 1972. The morning session was devoted to further testimony from BART management and statements from representatives of the major contractors for the system, which dealt mainly with the recommendations of the Post report which were concerned with the automatic train control system and the passenger safety issues. During the afternoon session, the Committee moved through the recommendations concerned primarily with administrative deficiencies. During the course of the hearing, testimony was heard from BART management, from the BART Board of Directors, from representatives of PBTB and from outside consultants, but not from the three dismissed engineers, nor from representatives of the California Society of Professional Engineers.

On Monday, February 5, 1973, Senator Alquist issued a statement following his receipt and analysis of the Blue Ribbon Committee Report:

> The report indicates that the BART automatic train control system as designed will not provide an adequate safety margin to permit full revenue operation. (underline added). However, the report also indicates a series of important recommendations that can make the system safe for the least cost to the taxpayer.
>
> In order that the system operate under full scale revenue conditions, the panel asserts that it is essential that critical train detection failures be eliminated by a combination of

> methods, all instances of train detection failures in the system must be automatically recorded, specific circuits must be redesigned, breaking tests under adverse conditions must be conducted, and circuit redundancy must be inserted where systems are vital to safety.
>
> In addition, the Panel recommended additional safety features that are regarded to make the system safe when operating in the mixed manual and automatic mode. These include such things as automatically entering all manual dispatching operations in the central computer before executing, providing a method for the operator to have greater control of the speed of the train when being operated manually, providing more information to the operator about the state of the system when he is operating a train, making provision for a true emergency stop, and altering the door design in order to reduce the hazard of dangerous door closing.
>
> Other important recommendations made by the Panel concerning failure to adequately provide for the safest meshing of the manual and automatic systems. <u>Suggested solutions center around more comprehensive ways to use BART's existing computer capability</u>. (underline added).

Backlash Within CSPE

As early as April 1972, evidence was appearing of a difference of opinion within the California Society of Professional Engineers on the BART case. During this month, when Gil Verdugo and Roy Anderson had driven to Sacramento from the Bay area for the meeting with Senator Nejedly, they stopped at the Sacramento airport to pick up Bill Jones. On the way from the airport to the Senator's office, Jones made it clear that if any attack were launched against the consulting engineers, he would have to withdraw his support.

Also in April, Ed Walker of the CSPE's Golden Gate Chapter expressed his reservations about the actions of the Diablo Chapter. Verdugo and Anderson described him at a dinner meeting they had with him:

> He was very nervous about any kind of investigation. He was telling us that BART had registered engineers that probably knew more about what was going on than those three engineers. The BART engineers were members of CSPE also. Finally he said, "You need to talk to the consultant engineers." And we said, "Fine we'd tried to talk to BART and they wouldn't talk to us. We were interested in talking to them but we hadn't found interest (on their part) so far." So he said, "Well, I'll set up a meeting and you can talk to them and get their views."

But again no meeting resulted.

Another indication of backlash in the CSPE was Bill Jones' action at the June 18, 1972 meeting in Max Blankenzee's home. Roy Anderson had prepared a report for Senator Nejedly and intended to discuss it at this meeting. But Jones was clearly opposed to any such action. Roy Anderson describes Bill Jones' attitude at this meeting:

> Bill was trying to back CSPE out of it and it wasn't clear why he was doing it, who had said something to him, or what.
>
> But it was apparent that he was really trying to get out of it. It was at that meeting that he told me not to give that report to Senator Nejedly. He said, "You're forbidden."

Perhaps Jones' position is explained by the following excerpt from an interview with him:

> Jones: And there was one of the members of CSPE who was active in this affair who rather quickly from the outset showed less interest in the problems of the three engineers as (sic) he did in hanging BART's dirty linen

Question: Any idea on which one that was?

Jones: That was Roy.

After the petition was made public, the backlash movement within the society gained in strength. On Monday, July 17, 1972, following the publication of the Diablo petition in the Sunday Contra Costa Times, Bill Bugge, BART Project Director for PBTB wrote a letter to Jim Walsh. The text of the letter is as follows:

Dear Jim:

Per our telephone conversation I am enclosing a copy of Contra Costa's Sunday Times relative to the action on the part of the Mount Diablo Chapter, California Society of Professional Engineers.

As discussed with you, I am concerned with their attack, particularly as it relates to PBTB. Statements such as, "Why must the taxpayer pay for all of the errors of the consulting engineers?" are pretty broad and to my knowledge, no effort has ever been made to determine the broad scope of the engineering that has been done on this project by Verdugo or his cohorts. It is extremely strange that a national professional society of engineers should allow a relatively small chapter of their organization to take this position without clearance from either the state organization or the national.

I appreciate very much your interest in this matter.

Very truly yours,

W. A. Bugge
Project Director

Following publication of the petition, Bob Kuntz, new President of the CSPE, called several members of the Diablo Chapter to meet with the legal counsel for the CSPE. The purpose of the meeting was to discuss the legal liabilities of the Society resulting from accusatory petitions. Kuntz describes his thinking relative to this meeting.

> They took it upon themselves to prepare and circulate a public petition. The petition called for an investigation. The public petition also made very, very direct accusatory statements as to, I guess one could say, what would be malfeasance on the part of the consulting firms involved in BART. As a result of that petition some of the consulting firms' engineers that belong to another chapter, the Golden Gate Chapter, became highly incensed because these were unsubstantiated, public accusatory statements. There is a section in the Code of Ethics that says engineers will not do that; you know, not falsely accuse and there are normal procedures for going about that. So they considered this to be malicious, unjustified, unethical, and even libelous. I had to call the Diablo Chapter officers to Sacramento to sit with legal counsel and to call their attention to the legal liability of circulating a public petition with accusatory statements of that nature. The legal counsel substantiated the concern and said that there should be a cease and desist. There should be no further actions like that because of the exposure that it would provide not only the Chapter but the State Society from a liability standpoint.

On Saturday, August 12, 1972, the Executive Committee of the California Society of Professional Engineers met. At that meeting, Bill Jones introduced a resolution to censure the Diablo Chapter. The following was unanimously approved:

> Exec. Comt. recommends that the Diablo Chapter be censured by the Board for attacking other engineers as they did in a recent petition at paragraphs three and four of that petition. Exec. Comt. further recommends Board ask the Diablo members cease to pursue such a course as contrary to NSPE Code of Ethics.

This meeting, which lasted from 9:30 a.m. to 4:30 p.m., was held in the conference room of Burns and Roe, Inc., Los Angeles, California. Attending the meeting and concurring in the resolution were Schonewille (First VP), Bill Schmidt, E. C. James, A. A. Dausch, C. G. Wilson, E. J. Honholt, Harry Lalor, Jim Walsh and Bill Jones. Guests at the meeting included Dr. Anthony, Ashley Martin, and Charles Yata. There was no representation at the Executive Committee meeting from the Diablo Chapter or from the East Bay Chapter.

On Saturday, August 19, 1972, the 25th Board of Directors meeting was called to order at 9:00 am. by President Kuntz in the San Francisco Hilton Inn, San Francisco, California. The last item in the Minutes reports the disposition of the Executive Committee recommendation:

> 6.00 NEW BUSINESS
>
> A. ...
>
> B. ...
>
> C. BART issue --
>
> 1. Executive Committee position stated "That the Diablo Chapter be censured by the board for attacking other engineers as they did in a recent petition at paragraph three and four of that petition. Further recommends that board ask the Diablo members cease to pursue such a course as contrary to NSPE Code of Ethics.
> 2. No formal action on this issue by the board at this meeting. Request that board receive copy of petition and copy of Golden Gate Chapter letter and that this be on agenda for next board meeting.
> 3. M/S/C/ to approve endorsement of Bill Holden, William Tarman, Billie Schmidt to serve as ad hoc committee of CSPE under guidance of Glenn Wilson and general control of Board of Directors to confer with three BART representatives on general technical subject matter.

These minutes were corrected as indicated in the minutes of the November Executive Committee meeting:

> to show the authorization of the *ad hoc* committee to investigate the ethical practice implications of 3 and 4 of Diablo Chapter petition relating to BART and committee to report to November Board meeting.

The Golden Gate Chapter letter referred to in the August 19 Board minutes was a letter ostensibly from Ed Walker to Bob Kuntz. The letter is dated August 17, 1972, and it was written on Golden Gate Chapter CSPE stationery. According to Ed Walker, the letter was actually written by Jim Walsh who cleared it with Walker and then signed Walker's name at the bottom. The letter asked why CSPE had not taken any action with respect to the petition circulated by the Diablo Chapter. "Walker's" letter expresses a concern over violations of sections 3, 4, 5, and 12 of the Code of Ethics. A sentence in the letter reads:

> I would request that you take this matter up at the next State Board Meeting which I understand will be held in the next few weeks.

A copy of this letter was directed to Gil Verdugo.

This letter was written after the Executive Committee motion to censure the Diablo Chapter and only two days prior to the meeting of the Board of Directors of the State Society. The published agenda for the Board meeting, however, had no indication of any agenda item on BART. Gil Verdugo describes the process by which he heard about the move to censure and the action that he took.

> I was at the Engineers in Private Practice Section Meeting (the night before the CSPE Board meeting) and someone said to me, "You're going to be censured." I said, "What the heck are you talking about?" And he said, "It's on the agenda."

According to Ed Walker, it was at the August 18 meeting of the Professional Engineers in Private Practice that Bill Jones recommended support of an action to censure the Diablo Chapter. It appears that this recommendation occurred prior to the time that Gil Verdugo came into the PEPP meeting.

Verdugo went to the Board meeting and he describes what happened when the motion to censure the Diablo chapter was considered:

> I stood up. I had to just stand they couldn't fail but to recognize me. I said, "The Diablo Chapter wasn't informed. We didn't have any prior notice of this sort of thing. We want to be heard before we're condemned. I want to talk to the judge and jury that is ready to hang us. Give us time so we can answer." So they backed off. They were ready to hang us on the spot.

According to Ed Walker, however, the Board of Directors meeting on the 19th ran out of time. Since the Board had to vacate the meeting room in the hotel, Verdugo simply did not have time to express his position.

Walker describes the Society as being split into two major camps over the BART issue:

> In one camp Roy Anderson was a very strong figure in this. Although both he and Gil Verdugo worked closely together, I feel that Roy was probably the strongest (sic) member. And I almost had a feeling of anti-BART in his thinking. Gil Verdugo I think was a little more reasonable about it. But I admit this, in all fairness of them, I did not have access to the volume of material that they were able to acquire.
>
> My own personal view, which seemed to be perhaps strongest in opposition to the direction they were going, was that there wasn't a fairness. It was all one-sided. You're taking one point of view and driving it very hard without giving the other registered professional engineers a chance to speak. And that hurt me...hurt my personal feelings of fairness toward the other side.

> At any rate, I guess that I would have to say that there were two rather distinct camps. I don't think that it was a true split. I think that it was one which did have some intense feelings at one time or another. We maintained our "cool" for the most part; we didn't really get hot-headed about it. It was a reasoning thing. I think that without Verdugo's rebuttal, that it (the motion to censure) would have been a recommendation for censure. Since Gil did not have a chance to finish his rebuttal, it's difficult to know how it would have turned.

CSPE President Bob Kuntz was responsible for appointing the <u>ad hoc</u> investigating committee called for at the August 19th Board meeting. He chose Trevor Smith, Eugene Sullivan, Howard Grant and Harold Williamson, and explains his choices:

> Trevor Smith has never been very active in the Society but he is a highly respected business man and professional in the San Jose Area. The other individuals were chosen because of their known support for the policies of the Society and because they were highly respected individuals. There was an attempt to pick people in the geographical area for several reasons: one was to minimize logistical problems, getting together and holding a hearing. The second was that this was the general area in which the situation existed and they were more attuned to actual situation, i.e. you know, the feelings of the people, the conditions that existed and so on. Howard Grant was a government employee, but he was also from the Golden Gate Chapter and also was a highly respected individual. There was an attempt to merge chapter, geographical, logistical, comprehensive behavior and also the government, private practice, and industry balance as best as possible.

The Ad Hoc Committee Reports

Officially Trevor Smith and Eugene Sullivan were co-chairman of the <u>ad hoc</u> committee; but in practice Smith acted as the chairman.

On October 13, 1972, Gil Verdugo received the October 11 notice announcing that the *ad hoc* committee would hold a hearing on Thursday, November 9, 1972 at 8:00 p.m. at the Engineer's Club of San Francisco. Admission to the hearing was specifically restricted to CSPE members. The hearing notice read in part:

> This hearing will be concerned *only with the methods* used in bringing the BART-D matter to *public* attention, and in particular the two statements discrediting "Consulting Engineers." in the Diablo petition. This Committee *will not investigate the merits of the issues presented in the* Diablo petition. The NSPE code of Ethics will provide a basis for the Committee's evaluation.

The hearing was well attended. Bill Jones and Ed Walker were two principal speakers in support of censure while Roy Anderson and Gil Verdugo were the two principal defenders of the Diablo Chapter petition. Many other CSPE members spoke, some favoring the move to censure and some defending the Diablo Chapter petition. The entire proceedings were tape recorded and the *ad hoc* committee adjourned the hearing just after 11:00 p.m. The Committee then reconvened in private to review and discuss the issues raised.

Their report, dated November 17, 1972, presents a brief review of the Committee activities and then gives a detailed discussion of Sections 2, 3, 4, 5, and 12 of the Code of Ethics as they relate to the specifics of their investigation on the petition. Several paragraphs of discussion are devoted to each Section of the Code of Ethics. The following sentences conclude the discussion of each Section:

> We do not feel that there has been a violation of Section 2.

> The Committee finds that there has not been a violation of Section 3.

The Committee finds therefore that there has been no violation of Section 4.

The Committee does not find that there has been a violation of Section 5.

The Committee finds that there has been no violation of Section 12.

It is the finding of this committee that action of the Diablo Chapter was in accord with the spirit of Section 2A of the Code of Ethics for Engineers.

Finally when each issue has been fully considered, the Committee reached a decision, and in each case that decision was unanimous.

The report was presented to the November 18, 1972 meeting of the Board of Directors and was adopted.

On November 20 Billie Schmidt wrote to Gil Verdugo. The text of his letter reads as follows:

Dear Gil:

At the Board of Directors meeting for the California Society of Professional Engineers on November 18, 1972 in Sacramento, California, your chapter, Diablo Chapter was commended for a job well done in its efforts to protect Safety, Health and Welfare with regard to the recognition of safety problems that exist in the Bay Area Rapid Transit District Control System.

This I believe vindicates your members in the Diablo Chapter activities which have presented problems under somewhat of a cloud.

There were many expressions of sincere appreciation given at the time that the attached resolution, brought forth by Ernie James, P.E. was moved, seconded and passed unanimously.

Let me personally commend and thank you and your chapter for an excellent job in safety.

Professionally yours,

B. A. Schmidt, P.E.

The attached resolution reads as follows:

RESOLUTION

<u>Whereas</u> The NSPE Code of Ethics states, "Section 2 - The Engineer will have proper regard for the safety, health, and welfare of the public in the performance of his professional duties. If this engineering judgment is overruled by nontechnical authority, he will clearly point out the consequences. He will notify the proper authority of any observed conditions which endanger public safety and health.

a. He will regard his duty to the public welfare as paramount."

<u>Whereas</u> The Diablo Chapter has recognized Safety problems existed in the Bay Area Rapid Transit District (BARTD) Control System.

<u>Whereas</u>... The Diablo Chapter, in its diligent effort to protect the public safety, health and welfare, regarding their duty to public welfare as paramount, conducted and reported on the alleged problems.

<u>Whereas</u> The report was brought to the attention of BARTD management with inconclusive results and subsequently to members of the State Legislature.

<u>Whereas</u> The Legislature requested the Legislative Analyst's Office to investigate the Diablo Chapter allega-

tions and found them substantially to be correct as documented in their report "Investigation of Operations of the Bay Area Rapid Transit District with particular reference to Safety and Contract Administration" dated November 9, 1972.

Whereas General Manager Stokes of the BARTD admitted with qualifications that the statements relative to safety in the Legislative Analyst's report were true but correctable.

Be it Resolved The CSPE commend the Diablo Chapter for a job well done in their efforts to protect the public safety, health and welfare.

The copy of the resolution that Schmidt enclosed with his letter to Verdugo had the following handwritten notation at the bottom of the page: "11-18/11 M/S/C at Board Meeting B. A. Schmidt." However, the official minutes of the CSPE Board of Directors meeting for November 18, 1972 make no mention of this resolution.

The National Society of Professional Engineers (NSPE)

The Society of Professional Engineers is basically a grass roots society. That is, most of the activity of the organization is at the chapter level. Local chapters are banded together to form state societies to deal with those issues which are appropriate for statewide considerations. State societies, in turn, are banded together to form the national society to permit them to deal with issues on a national level. While it is true that individual members are members of the national society as well as their state societies and local chapters, the operation of the organization is much more a hierarchy of confederations than a monolithic national organization.

All members of the California Society of Professional Engineers interviewed in this study agree that there was no involvement in the BART case at the NSPE level. But even though the national society took no active role in attempting to achieve a reconciliation of the three engineers with the BART organization, it was very much aware of the activities taking place in California. On September 13, 1972, for example, Robert L. Nichols, at that time the NSPE Chairman of the Professional Engineers in Private Practice, wrote a letter to the Diablo Chapter. He expressed his distress over the Chapter's activities in the following way,

> This matter was discussed at our recent NSPE/PEPP Executive Board meeting with considerable concern. It is very distressing that a Chapter within NSPE has seen fit to raise such serious questions concerning an area of practice of engineering.

A copy of that letter was sent to Bob Kuntz as the current President of CSPE.

Perhaps the best way to understand the nature of the interaction between NSPE and CSPE is to look at an exchange of letters which took place between James F. Shivler, Jr., the NSPE President, and Robert J. Kuntz, the CSPE President.

> September 22, 1972
>
> Dear Bob:
>
> This will confirm our telephone conversation on September 18 relative to two matters of concern which have been brought to my attention.
>
> The first involves the action of CSPE's Diablo Chapter in preparing and distributing a petition to the California State Legislature to conduct a "BARTD Investigation." This item has, I believe, been brought to your attention by copy of a letter from Bob Nichols, to the Diablo Chapter. You advised me that this Chapter

action was taken without the knowledge or concurrence of the leadership of CSPE. The propriety of this action has been seriously questioned by CSPE and it is 'my understanding that your Executive Committee has recommended that those responsible be censured and that this matter will come up for further action at the next CSPE Board of Director's meeting.

The second item concerns a letter prepared by W. Schroeder, P.E. Chairman, CSPE Economic & Representation Committee, under a CSPE letterhead, addressed "Dear Fellow Engineer" and dealing with the "affiliation agreement" of the CSPE with the Western Council of Engineers. This letter is accompanied by literature of the Western Council of Engineers, including an "authorization card" whereby an interested engineer could designate WCE to represent him as "sole Collective bargaining representative." You advised me that you were unaware of the Schroeder letter and based on my representation of its content expressed serious concern over the implications of such a procedure. It is my understanding that you will thoroughly investigate this matter to determine all of the facts.

I am sure that you are as concerned as I over both of the matters discussed above. It is my understanding that you will let me have a full report as expeditiously as possible.

Personal regards.

Sincerely,

James F. Shivler, Jr., P.E.
President

In response, Shivler received the following letter from Kuntz:

October 11, 1972

Dear Jim:

I promised that I would provide you with some information regarding the petition that had been circulated by the Diablo chapter covering the BART issue. You will recall that two items in the petition were statements mentioning the field of consulting engineering. Several pieces of correspondence have crossed my desk indicating the concern of some of our NSPE leaders for the seemingly derogative accusations. The following actions have been taken:

1. An ad hoc committee has been established to hold hearings concerning the ethics of such accusations made in the public petition form.

2. The Diablo Chapter has been strongly counseled that this type of activity does not provide the type of reaction needed to address the basic problem with BART and the improper industrial relations policies indicated by the firing of the three BART engineers.

3. I personally communicated, to the president of the Diablo Chapter, my displeasure with two of the items in the subject petition. Furthermore the CSPE legal counsel mildly took to task the chapter president because of the potential of libel litigation--as remote as that possibility is.

This subject has been extensively discussed by the executive committee of CSPE, and a motion of censure was proposed to the CSPE board of directors at its August meeting. No action was taken by the board because of a time limitation. However, this subject is part of the agenda for the November board meeting at which time the ad hoc committee will present its findings. The executive committee has requested that the ad hoc committee explore the ethics involved in this particular chapter's action. Furthermore, the executive committee plans to request the NSPE Board of Ethical Review to examine the apparent dichotomy between two sections of the Code of Ethics. The Diablo Chapter has taken the position that it was their ethical responsibility to publicize major deficiencies in the engineering approach of the BART system, since the engineering Code of Ethics demands that public health and safety be over-riding in any issue. They felt that illumination of deficiencies in this area was mandatory to achieve the legislative investigation necessary to correct a system which is hazardous to public welfare. As I am sure you are now aware there has been a malfunction of the system similar to that which was predicted by the three engineers who were discharged by BART for their disclosure. The accident that occurred caused the injury of five passengers but could have caused many fatalities just as well. The Diablo Chapter agrees that better wording could have been used in their petition, but felt that their behavior is in concert with the highest principles of engineering ethics. I might say that many members of the Executive Committee share the feelings of the Diablo Chapter.

CSPE has been involved in the BART issue since its inception. Deficiencies in the cost estimates made by the original consulting engineering firm were identified and the 100% over-run currently being experienced was predicted. The question to be resolved is, to what extent must an investigation be made to identify a deficiency in engineering before public disclosure of this deficiency is ethically permissible? As I mentioned, the CSPE ad hoc committee will investigate the ethics of this particular issue but we strongly recommend that the NSPE Board of Ethical Review be directed to examine the general subject. With engineering being responsible for the technical, fiscal, and social applications of science and technology, the engineering Code of Ethics may have to be examined in light of the strong implication of over-riding social and environmental issues in any engineering project.

I will keep you informed as to the results of the ad hoc committee. The Diablo Chapter has agreed to cease any further action on the controversial petition. If I can be of any further help, please let me know.

Yours in the profession,

Robert J. Kuntz, P.E.
President

A second letter from Kuntz to Shivler was written the following month.

November 28, 1972

Dear Jim:

Enclosed you will find a copy of the report from our *Ad Hoc* Committee on Ethics - BART. You will recall that there was some concern expressed by yourself and Bob Nichols with the action of our Diablo Chapter in connection with the circulation of a public petition asking for a legislative investigation of BART. The enclosed report by our *Ad Hoc* Committee is one of the most comprehensive studies conducted on a specific issue of this nature that I have been in our state society. You will note that the findings of the *Ad Hoc* Committee "clear" any concern for the ethics question associated with the Diablo Chapter actions.

The intention of the Diablo Chapter was to bring to public notice the deficiencies of the BART system which had already caused the injuries of many individuals. The efforts reached fruition recently with the conclusion of the legislative investigation for which the Diablo Chapter was given credit by the California Legislative Analysts. The report validates the many areas of deficiencies in the BART system which have become a matter of public safety.

The enclosed report from our *Ad Hoc* Committee was submitted to our Board of Directors on November 10th and received unanimous approval. No further action with respect to the "ethics" of the Diablo Chapter was planned. However, we feel strongly that the matters of ethics should be resolved and those which have so severely damaged the careers of the three engineers who, through their dedication to the public safety, have lost their jobs and have been blacklisted. We shall continue to pursue every avenue possible to achieve retribution for those actions taken by a public agency. This may indeed be an excellent cause to which to apply some of the NSPE legal action fund.

If I can be of any further assistance in providing further information concerning this highly visible subject, don't hesitate to contact me.

Yours in the profession,

Robert J. Kuntz, P.E.
President

The treatment of BART and the BART case in the pages of the NSPE monthly publication, Professional Engineer, suggests something of the attitude of the national organization toward the issues. In August, 1972, Roy Anderson had a luncheon meeting with Milton Lunch, the NSPE attorney, and John Kane, the editor of Professional Engineer. Mr. Kane expressed interest in the BART case and urged Anderson to submit an article for publication. Anderson, therefore, prepared a manuscript which he sent to Kane on October 9, 1972. But the manuscript was never published.

On the other hand, the September issue of Professional Engineer carried a series of three articles on BART. The first article, "Engineering for BART", was written by William A. Bugge, the Project Director for PBTB. The second, "BART's Economic Spinoff in the Bay Area" was written by Dave Hammond, BART Assistant General Manager for Operations and Engineering, and the final article in the series was a Westinghouse technical report entitled, "Computer Supervision of BART's Transit Control System." Needless to say, all three articles presented a very positive view of the BART system development. No mention was made of three engineers being fired nor of any safety problems associated with the automatic train control system.

The February, 1973, issue of Professional Engineer contained the first mention of the BART case. Buried on page 73 among five pages of "NSPE News/Impact" items was a 260 word item entitled "Diablo Chapter's BART Report Starts State Investigation."

The only other mention of the BART case in the NSPE Journal is in conjunction with the awarding of the NSPE Journalism prize to Justin Roberts as reported in the June, 1973 issue.

CSPE Fades Away

By the middle of November, 1972, the level of support within the CSPE for the case of the three engineers had undergone an almost unbelievable oscillation. In March, April and May, the dominant majority of vocal members within CSPE strongly supported the engineers and backed CSPE involvement in the case. In June, July, and August, the position of support came under strong criticism and the Diablo Chapter was threatened with State Society censure for its actions in the case. The _ad hoc_ committee hearing on November 9 provided a forum for a frank discussion of the ethical aspects of the petition sponsored by the Diablo Chapter. As a result of these and other deliberations, the _ad hoc_ committee offered a resounding statement of support for the Diablo Chapter and consequently for continued CSPE support of the three engineers.

In spite of the _ad hoc_ committee's strong statement of support, however, the level and the effectiveness of the CSPE support for the three engineers declined dramatically in the later part of 1972. The single factor that seems most responsible or this decline in activity was the departure of Roy Anderson from California to a new job with the Transportation Safety Board in Washington D.C. in late November, 1972.

Bill Jones, however, feels that CSPE lost its grasp on the matter of the BART engineers shortly after the May meeting in 1972. He comments:

> "I guess it must have been some time in June of '72. CSPE completely lost its grasp. The

> three (engineers) were wronged. If they were going to proceed with any kind of action, they had to fund it themselves. They started going their own merry way...BART just disappeared out of sight. Or at least those three engineers disappeared out of our sight. CSPE really lost track of this. The only person who, from time to time, would remind us that there's still three engineers out there who got treated badly by BART was Gil Verdugo.

When Jones was asked whether the CSPE attitude could be described as lethargic, he responded,

> I think lethargy is not quite the right word. There was inertia in the system. To a large extent based on lack of any specific knowledge of what was happening and what we could do with very limited funds. We always come back to that, limited funds.

The February 10, 1973 minutes of the CSPE Executive Committee accurately reflect the prevailing attitude at this time.

> The Executive Committee requests the board to reaffirm its position in support of the gentlemen released from BART. Executive Committee designated Dr. Bruce Anthony as a spokesman (counsel and guidance provided by Ed Walker, Bill Schmidt, Jim Walsh, Gil Verdugo and the released men) to continue quiet, formal negotiations with BART decision makers with all due speed. The charge is to resolve the entire matter to the satisfaction of all parties. This gesture is in behalf of the released engineers. It is not a matter between BART and CSPE or BART and the Diablo Chapter. It is a matter between the engineers who were released and their former employers.
>
> A resolution was submitted by the Diablo Chapter and modified by the Executive Committee for the board's adoption.

On September 17, 1973, the board adopted the following resolution,

WHEREAS the Constitution of the California Society of Professional Engineers state that, "The objectives of the Society shall be the advancement of the public welfare and the promotion of the professional, social and economic interests of the Professional Engineer", and

WHEREAS the California Society of Professional Engineers came to the defense of members of the Engineering Profession and the interests of the general public in the San Francisco Bay Area Rapid Transit issue and

WHEREAS to date BART has failed to fully respond to this Society's efforts to mediate a satisfactory settlement, in behalf of three engineers fired by BART, and

WHEREAS the engineers acted in the interest of public welfare, NOW THEREFORE BE IT RESOLVED that the Society continue its support of legal action funds for the three engineers and actively coordinate action in their behalf.

CSPE as a Friend of the Court

A year and a half later, on September 24, 1974, Leendert Schonewille, successor to Robert Kuntz as President of CSPE, wrote to Harry Lalor,

"Dear Harry:

Subject: BART matter - URGENT

We have now received a formal invitation to provide a "friend of the court" brief in cooperation with the plaintiffs Bruder, Blankenzee, and Hjortsvang.

A San Francisco attorney will be reviewing the case and will assess the appropriateness of our filing the brief. We hope he can meet with us Monday evening.

Harry Lalor will call an emergency meeting of the committee I have appointed: Harry Lalor,

> John Haff, Billie Schmidt, Gil Verdugo, Ed Walker, and Hal Williamson.

This committee met for the first time on Monday evening, September 16 at 7:00 for a dinner meeting at the Hilton Inn, International Airport, in San Francisco. The consensus of this committee meeting was expressed in a letter from Lalor:

> CSPE should write a friend of the court brief admitting ignorance as to the acts in the case; however, requesting the court to consider the ethical question of the individual engineers responsibility for public health, safety and welfare in its deliberations.
>
> It was felt that if such a brief was not acceptable to the court that alternative one (do nothing at this time) would prevail.
>
> Harry Lalor was directed to obtain advice from local counsel as to whether or not this recommendation was feasible.

Two days later, Harry Lalor met with Chuck Luckhardt, attorney-at-law, San Jose. Following this meeting, Lalor concluded that the CSPE should proceed with the petition to the court to file the brief.

On November 26, 1974, Harry Lalor advised the BART Brief Ad Hoc Committee members that Charles Luckhardt, the BART Brief Attorney, doubted that CSPE would be allowed to file an opinion in the BART case. Nevertheless, the committee members were asked to research the matters individually and to provide, on or before December 20, 1974 any opinions, references, and/or recommendations as to what the brief should say. The committee members responded with a variety of opinions and points of view. Finally, on February 5, 1975, Harry Lalor wrote to Bob Kuntz who by this time had taken the position of Executive Director of the CSPE.

Dear Bob:

Enclosed for your information and files is a brief prepared by IEEE in conjunction with the "BART" case.

While the verbage appears to be somewhat self-serving, it is a factual and professional approach that would have been worthy of our co-endorsement.

This copy was forwarded to me by Charles Luckhardt our Attorney in this matter, with the recommendation that CSPE join in this brief. However, this point is now mute as it is my understanding that the case has been settled out of court.

Very truly yours,

Harry N. Lalor

The Price of Public Involvement in Behalf of the BART Engineers

The three individuals most deeply involved in the activities of the CSPE related to the BART case were Bill Jones, Gil Verdugo, and Roy Anderson. What was the effect of their involvement on their personal and professional lives?

In March of 1972, Bill Jones was President and Chief Engineer of Gribaldo, Jones & Associates, Mountainview, California. When asked what his personal or professional losses or gains were as a result of his active participation in this case, Jones responded that he felt he had suffered a business loss. He comments,

> And I had a personal loss because I spent so much time on the case and other things like this that I came to a parting of the ways with my old company. I wasn't fired or anything like that. No, I was the president of the com-

> pany and major stockholder. But it was definitely a factor in grumbling to the fact that I wasn't getting as much work done as was otherwise expected of me. This led to a situation in which I felt it might be easier for me to cut out than to try to fight it; so I just cut out...I had people at work ask me, or almost tell me, not to be involved in this particular BART situation because it wouldn't do well for the company in its attempt to develop relations with the consulting engineer's firms.

In 1973, Bill Jones formed a new company, William F. Jones, Inc.

In March of 1972, Gil Verdugo was also in private practice. He wonders whether his involvement in the BART case was responsible for a decline in his business.

> I've often wondered...I think we were caught up in the economy slump and then the environmentalists pursued this stopping of all sorts of development so it (the fall-off in his business) might have been a coincidence. But once in a while I wonder, you know, whether some of those guys working for the larger consulting firms didn't spread the word around a little bit, that it was bad news to get involved with us. I don't know. That part of it bothered me. But it doesn't bother me anymore. Funny little things would happen once in a while. I ran across some people who'd say, "Oh, hey you're the guy who tooted the horn on those things. I heard about you." What did you hear? And then it was quiet, they didn't talk about it anymore. Roy and I discussed this several times. Hell, we weren't afraid to face anybody, and we could hold our heads up. I wasn't ashamed of any of it.

In March of 1972, Roy Anderson was working for the Division of Bay Toll Crossings, part of the California State Government structure. Shortly after Anderson spoke publicly on the BART affair, the Chief Engineer of his branch called to ask him what was going on. Anderson responded by asking whether or not he was being asked to keep quiet. He recalls,

> They were very diplomatic about it, they said things like, well, we'll leave it up to you, but use some discretion and try not to get our agency involved. What you say is, Roy Anderson says...don't be a spokesman for our agency.

In spite of any business or personal losses experienced as a result of their active involvement in this case, all three men are adamant in their belief that what they did was proper, correct, and professionally responsible. Each of them would, if given the opportunity to relive this period of time, again play an active role in the drama. Perhaps Roy Anderson says its the best:

> I wouldn't trade the experience for anything in the world. I think it's the most interesting, fun thing I ever did in terms of the gravity. I just enjoyed tremendously the opportunity of doing what I did. It's something you don't often get the opportunity to do.

Jones comments:

> I wouldn't change a thing I did. Not a thing. I think the matter was pursued as best could be done under the circumstances. The only thing I might have done is to bear down a bit more strongly on the BART management and consultants in the period in March, 1972, when I was trying to arrange a meeting with them before we went to our lawyer.

Finally, Gil Verdugo says:

> Personally and from a public standpoint, it was gratifying. I got letters from people saying, "Gee whiz, it's about time some engineers stood up and it's nice to see that your group came to the defense of those three engineers." Some of the other comments that were made, "Boy, it's sure refreshing to see and hear people speaking up on issues like this."

Stephen Unger Becomes Interested

Steve Unger is a professor of computer engineering in the Department of Electrical Engineering and Computer Science at Columbia University in New York City. At just about the time that Hjortsvang, Bruder, and Blankenzee were getting into trouble with BART, Steve Unger became involved in an issue of professional employment practice in New York.

Sometime in 1971 or 1972, Larry Tate, an engineer with IBM, was summarily fired from the company. The problem started with a small party which Tate's teenage daughter held at her home for a group of friends. The police raided the party and found some marijuana. Tate got into an altercation with the policemen and was arrested. Immediately after his arrest, IBM fired him. Tate had friends among the computer science faculty at Columbia who promptly drew up a petition of public support for Tate and against the firing by IBM. Steve Unger was instrumental in the petition drive and in the effort to bring pressure to bear on IBM. The efforts were successful and Tate was rehired by IBM. The criminal case against Tate was ultimately dropped and all charges against him removed from the record. The experience--and his success--had increased Unger's interest in codes of ethics and guidelines for professional employees and their employers.

Unger's interest in BART and in the BART case was stimulated by a series of articles, editorials and letters which appeared in _Spectrum_, the publication of the Institute of Electrical and Electronic Engineers.

In the fall of 1972, Gordon Friedlander, a senior staff writer for _Spectrum_, began a series of articles on BART. "The BART Chronicle", September 1972, presented the history of the BART system and pointed out some of the problems that BART was having. "BART's Hardware--from Bolts to Computers", October, 1972 presented in some detail the technology that was under development for the BART system. "More BART Hardware", November, 1972 concluded this series with the presentation of some final details of the system design and a look at the aesthetics of the system. In general, all three of the articles were extremely complimentary of the introduction of advanced technology into a modern mass transit system. A discordant note, albeit a weak one was struck in *Spectrum's* November editorial. Entitled "Bugs in BART", the editorial described a round trip on the BART system which, by and large, was a positive experience but which concluded with a brief discussion of the Fremont Station accident that had occurred 20 days after BART's official opening on September 11, 1972. The editorial identified the cause of the accident as a malfunction in the crystal oscillator of the automatic train control system on board the lead car of the train.

The December 1972 issue of *Spectrum* contained two contributed letters. One, critical of the system, was called "BART Reliability" and was contributed by Holger Hjortsvang. The other was a rebuttal of Hjortsvang's points by Bill Bugge, BART Project Manager for PBTB. Hjortsvang's letter made no mention of his being fired from BART; he referred to himself only as a "former systems engineer for BART." Neither did the letter from Bugge indicate that Hjortsvang had been fired.

The problems in the BART system were given extensive coverage in two further articles written by Gordon Friedlander for the March 1973 and April 1973 issues of *Spectrum*, "Bigger Bugs in BART?" and "A Prescription for BART." The March article began with a discussion of the Fremont Station accident and went on to mention the Post Report, the Battelle Institute Report, the comments by Dr. Wattenburg, the California Public Utilities Commission hearings, and the blue ribbon panel report. The April article continued the discussion of the problems inherent in BART and reported the recommendations that had been made for improving the overall system. Particularly, the safety of the system and, most particularly, the safety the automatic train control system were discussed. Friedlander's two articles in 1973 do not mention the three engineers who were fired from BART. But in the research that Friedlander did to write the article he did communicate with Max Blankenzee.

When the March and April Spectrum articles appeared, Unger was motivated to learn more about the problems inherent in BART. He was interested in transportation and he was attracted to and intrigued by the idea of having high level technology employed in a mass transit system. He was outraged, however, at the apparent ineptness with which the technology was being implemented--at least according to Friedlander's reports. On May 2, 1973, Unger wrote to Friedlander asking for more information and for additional sources of information on the BART system. Friedlander replied with a long list of names and copies of a few of the documents he had used as source material in the preparation of the articles. One of these documents had a particularly strong effect on Unger. This was a letter from Bernard Oliver, a member of the blue ribbon investigating committee appointed by the California Senate Committee on Public Utilities and Corporations. The letter was addressed to Dr. Cy Herwald, Vice-President of Westinghouse Electric. The two men had known each other for a number of years and each had served as president of the IEEE. The three-page letter discussed some of the details of the automatic train control systems. Two sentences from this letter give an indication of its overall tone.

> There are other aspects of the train control system that suggest the design did not enjoy the attention of your top people. In some instances, it is not clear whether the fault was in the specifications or in their interpretation, but the result was the same: BART did not get the best system modern technology could provide for the price that was bid.

This letter, plus other materials received from Friedlander spurred Unger to write a spate of letters. Records indicate that he wrote some 25 letters during May, June and July of 1973 to such persons as Hjortsvang, Bruder and Blankenzee, (the engineers), Stokes, Fendel, Kramer, Follett, and Wargin and Profet (of BART Management), Wattenburg, Helix, Verdugo, Anderson and Justin Roberts (CSPE members and other interested parties). Unger collected his information and wrote an article, "The BART Case: Ethics and the Employed Engineer", which was published in the September, 1973 issue of the Newsletter of the IEEE Committee on Social Implications of Technology.

To understand Unger's relationship to the IEEE organization and the impact that his information gathering and dissemination had on the organization, we need to step back from the details of the BART case and look briefly at the changes which were underway in the IEEE in 1971, 1972, and 1973.

Winds of Change in the IEEE

In the spring of 1971, the IEEE Board of Directors received a petition for amendment of the organization's Constitution which would make its primary purpose the promotion and improvement of the economic well being of the membership and would give a secondary role to its scientific, literary and educational activities.

The proposed Constitutional amendments appeared on the 1971 ballot. The official position of the Board of Directors was to recommend against their adoption, and the amendments failed to secure the necessary 2/3 majority of those voting. But of those voting (about 35% of those eligible), a slight majority of the United States membership (52%) was in favor of the proposal; of the non-USA IEEE membership, however, a clear majority (71%) opposed the proposed amendments.

The petition action and the relatively substantial support that it received among USA members led the Board of Directors to attempt to assess more accurately the desires of the membership relative to the non-technical pursuits of the Institute. The original Constitution of the organization, approved in the 1963 merger of the American Institute of Electrical Engineers and the Institute of Radio Engineers, delineated the purpose of the organization as,

> scientific and educational, directed toward the advancement of the theory and practice of electrical engineering, electronics, radio and the allied branches of engineering and the related arts and sciences; means to these ends include, but are not limited to, the holding of meetings for the reading and discussion of professional papers, and the publication and circulation of works of literature, science, and art pertaining thereto.

Thus, by nature of the Constitution and by nature of the attendant Federal Internal Revenue Service classification, the Institute had to limit in its non-technical activities to an insubstantial fraction of its total efforts.

In December, 1971, the Board submitted to 145,000 USA members of the Institute a questionnaire which asked for opinions relative to possible activity of the Institute in non-technical areas. Over 57,000 returns showed a greater than 2 to 1 majority in favor of increased non-technical activity on the part of IEEE.

The Executive Director of the Institute, Donald Fink, was then directed to study the possibilities of Constitutional amendments and to prepare appropriate drafts. Ultimately a Constitutional amendment was proposed to the membership for their ratification. The June, 1972 issue of *Spectrum*, IEEE's monthly publication, carried an extensive article which explained the historical development of the Constitutional amendment question and presented the proposed Constitutional changes to the membership for the first time. It was proposed that the purposes of the Institute be broadened by appending to those in the original Constitution the following additional ones:

> Professional, directed toward the advancement of the standing of the members of the professions it serves; means to this end included, but are not limited to, the conduct and publication of surveys and reports on matters of professional concern to the members of such professions, collaboration with public bodies and with other societies for the benefit of engineering professions as a whole, and the establishment of standards of qualification and ethical conduct. The IEEE shall not engage in collective bargaining on such matters as salaries, wages, benefits, and working conditions customarily dealt with by labor unions.

> The IEEE shall strive to enhance the quality of life for all people through the world through the constructive application of technology in its fields of competence. It shall endeavor to promote understanding of the influence of such technology and the public welfare.

With this proposed change, the non-technical concerns of the Institute could be given at least as much weight as its technical activities. The explanation accompanying the proposed Constitutional changes indicated that the key word in the proposed change was "professional" with the qualifying clause "directed toward the advancement of the standing of members of the professions it serves."

Simultaneous with this constitutional move to enable wider activity in non-technical areas, the Board of Directors, in February 1972, received a petition signed by more than 600 members, senior members and fellows of the IEEE requesting that a Committee on the Social Implications of Technology (CSIT) be formed. The Board of Directors received the petition and decided an _ad hoc_ committee on the Technical Activities Board (TAB) was more appropriate than a professional group, and an organizational meeting was called for June 24, 1972. In December, 1972, the first issue of the CSIT Newsletter appeared. Subsequent issues appeared, initially, on a quarterly basis.

The December, 1972 issue of Spectrum gave the following results of the vote on the Constitutional amendment question: 42,899 in favor, 6,508 against, 1,543 invalid returns. The lead paragraph in the Spectrum article reads:

> The just-tallied vote of IEEE members constitutes an important landmark in a rapid but well considered series of actions undertaken by IEEE that will add new dimensions to an already important function of the Institute. That function represents professional activities such as employment practices and guidelines, pension plans, interaction with government at all levels, and man power planning, among others, and the vote was overwhelmingly in favor of amending the Constitution of IEEE to engage in matters of legislative, social, ethical, and economic concern to the membership.

Even before this vote was in, the Board of Directors had moved forward in this professional area and established the United States Activities Committee (USAC), consisting of six Regional Directors and a Vice-President, Regional Activities. The Board also created an office in Washington D.C. to provide technical liaison with Congress and with the administration.

Thus, the organizational climate within the IEEE was prepared for the largest technical professional society in the world--180,000 members strong--to set sail into the new and uncharted waters of professional activities for the benefit of its members.

One of the problems in such a large and voluntary organization is that of internal communications; people in one part of the organization don't know what those in another part are doing. Typical of the communications problems within the IEEE was the designation of the Committee on the Social Implications of Technology as an arm of the Technical Activities Board at almost the same time that USAC (later to be renamed United States Activities Board (USAB)) was created and placed at an equal level in the organization chart with the Technical Activities Board. Among USAC's subcommittee was a Committee on Employment Practices, the name of which was later changed to the Committee on Ethics and Employment Practices.

It appears that the IEEE considered the case of the three BART engineers in the following fashion: The issue attracted the attention of Steve Unger as an individual. In his role as Chairman of the Working Group on Ethics, Unger brought the case to the attention of the Committee on Social Implications of Technology. A resolution of support for the three engineers was drafted, approved by CSIT, passed up to the Technical Activities Board, considered, modified, approved, and sent on up the organization to the Board of Directors. The Board of Directors then referred it to the United States Activities Committee which in turn referred it to the Committee on Ethics and Employment Practices. A slightly different resolution then came from the Committee on Employment Practices back to USAC and back to the Board of Directors. At almost exactly the same time these activities were taking place, the Board of Directors appointed its own _ad hoc_ committee to study the case. Ultimately, through both channels, the whole issue arrived back at the Board of Directors for their action. The detailing of this complex flow in the paragraphs below is intended to show some of the complexities and difficulties attendant on professional society involvement in cases of ethical conflicts between employed engineers and their employers.

The CSIT Approaches the Board of Directors

On May 21, 1973, Steve Unger wrote an open letter of invitation to IEEE members to join the working group on ethics. In that letter he indicated that he was in the process of gathering information on the BART case. A month later, on June 27, 1973, the Committee on Social Implications of Technology instructed Unger to continue his investigation of the BART case and to bring recommendations to the Committee for possible action of the Institute. Unger reported the results of his investigation in the September 1973 issue of the CSIT Newsletter. Then, on November 10, 1973, the Committee on Social Implications of Technology met in New York City and heard Unger's report on a broad range of activities associated with his chairmanship of the Working Group on Ethics. He reported on his participation on the Subcommittee on Ethics of the USAC Employment Guidelines Committee, and on the current draft of a simplified text of a code of ethics, and added his view that a need existed for a new IEEE mechanism to support the ethical engineer. He reported further on two conferences that dealt, in part, with ethical matters, and finally he informed the Committee on the status of the BART case. Unger was asked by the Committee to obtain information on how to start a defense fund led by a prominent and professionally invulnerable engineer and buttressed by impeccable financial accounting. The purpose of such a fund would be to demonstrate professional support rather than to provide for sustenance and for legal expenses. Dr. Unger recommended continuing coverage of the BART case in the CSIT Newsletter.

On March 25, 1974, CSIT held a open meeting in New York City and the following resolution was passed unanimously (with 16 affirmative votes):

> Whereas in the practice of their profession, employee engineers sometimes face conflicts between what they perceive to be the public interest, health, or safety and the demands of management. They sometimes face reprisals from their employers if they act in conformity with professional ethics. A notable case in point is that of the three BART engineers, as described in the attached article.

> Resolved that the Committee on Social Implications of Technology hereby request:
>
> a) That the IEEE establish mechanisms for providing support to engineers whose acts in conformity with ethical principles may thus have placed them in jeopardy. These procedures could include information mediation, formal investigation followed by publication of the facts, litigation, and public condemnation of unfair employer practices;
>
> b) That even before eventual establishment of such procedures, the IEEE intervene in the case of and in support of the three BART engineers to help establish an important precedent for the engineering profession.
>
> CSIT is prepared to designate one or more representatives to present the facts of the BART case to the IEEE Board of Directors.

Although this resolution was approved by CSIT for transmittal to the Board of Directors, it had first to go through the Technical Activities Board, since CSIT was a committee of that group. On May 17, 1974, the resolution was considered by TAB. Although the minutes of the meeting indicate that the resolution was modified, passed, and then sent to the Board, our research indicates that the resolution was transmitted to the Board without modification.

From the Board to USAC and the Ad Hoc Committee

On May 18 and 19, 1974, the CSIT resolution was received and discussed by the Executive Committee of IEEE's Board of Directors. During the discussion, the question was raised as to whether the three BART engineers were themselves entirely ethical in their behavior, and the possible cost to IEEE of involvement in the case was tentatively explored. Fears were expressed that active support of the engineers could bankrupt the IEEE. The Committee finally instructed IEEE President John Guarrera to form a task force to commend a policy to be followed by IEEE on support of ethical engineers.

Guarrera appointed Bob Saunders as chairman to serve with Frank Barnes, Hal Goldberg, Bob Shuffler, and Leo Young. All were Directors of the IEEE. The charge to the *ad hoc* policy committee was:

> To develop policy recommendations for the Executive Committee and Board of Directors concerning the manner in which the IEEE may act on behalf of those of the profession who may suffer loss, financial or otherwise, as a result of positions taken on matters of an ethical nature.

Guarrera also referred the CSIT resolution to the USAC Ethics and Employment Practices Committee.

Following this Executive Committee meeting, Donald Fink, the Executive Director of IEEE, had several discussions with the IEEE counsel. These discussions led to the concept of an IEEE involvement in the BART case in the form of an *amicus curiae* (friend of the court) brief in behalf of the engineers. The concept of an IEEE *amicus curiae* was then so strongly associated with the CSIT resolution that the August 1-2, 1974 minutes of the Executive Committee meeting note that "The CSIT request for an *amicus* in support of BART engineers was referred to the *ad hoc* policy committee and to the USAC Ethics and Employment Practices Committee." In contradiction to these minutes, our evidence suggests that the concept of an *amicus curiae* in support of the engineers originated, not with CSIT and not with the CSIT resolution, but rather with conversations between Donald Fink and the IEEE legal counsel.

Someone from the Ethics subcommittee of the USAC Committee on Ethics and Employment Practices contacted Steven Unger during the summer of 1974 for information on the BART case. Unger provided his entire file of documents to be copied. In addition, members of the Ethics subcommittee contacted a number of informed persons on the West coast to gather relevant information first hand. Following review of all these documents, the subcommittee concluded that IEEE should support the three engineers, and a recommendation to that effect was made at the September 8, 1974 Ethics and Employment Practices Committee meeting. Again questions concerning the ethical behavior of the three engineers were raised and the issue was sent back to subcommittee for further study.

On October 26, 1974 Zourides, Sassanaro, Snyder, and Cummings, the USAC legal counsel, met at the Holiday Inn at La Guardia Airport in New York City. This *ad hoc* sub-subcommittee of the Ethics and Employment Practices Committee struggled with the problem of defining a proper role for the IEEE in the BART case. Their meeting began at 10 in the morning and, after several hours of discussion, the new idea of an *amicus curiae* brief based on defining a proper standard of ethical behavior for an employed engineer developed. This concept of the brief, based on defining an ethical standard was opposed to one simply in support of the position of the three engineers, was seen as the perfect solution to the problem at hand. The IEEE would not be viewed as taking sides in the conflict between employer and employee; it would, however, set a standard for proper ethical behavior and would leave to the court, with its superior ability to gather the facts, the determination as to whether the engineers had met or had violated the standard of ethical behavior. This course posed only small financial and professional costs and risks to the Institute. By mid-afternoon that same day, Joel Snyder and Frank Cummings were on their way to Washington to present the solution to the USAC Steering Committee. The idea was immediately accepted and ordered to be prepared for USAC and for the Board of Directors meetings to be held in December, 1974.

On November 2-3, 1974, IEEE Executive Committee met and Mr. Stern (the president-elect of the Institute) moved that,

> with respect to the BART case, the Executive Committee recommend to the Board of Directors the adoption of a friend of the court position, with the justification that our members are expecting us to take a position on matters of this type, that this would be a contribution to increasing the professional activities and involvement of the Institute on an objective basis and that it would be a public service. (unanimously approved.)

In December, 1974, Dr. Saunders reported for the *ad hoc* policy committee to the Board of Directors. He reported that the *ad hoc* committee had met three times to respond to its charge. The findings of his committee were reported as:

1. It is appropriate or the IEEE to enter a brief, as recommended by USAC, as an amicus curiae in specific cases upon the authority of the Executive Committee. It is also appropriate for the IEEE to publicize its actions in a suitable manner so that the members and the public at large are aware of the responsibilities of the engineer toward the public safety and interests. It is not appropriate for the IEEE to assume an adversary position or intervene in any court action on behalf of or against an individual member of the Institute in a matter of ethical principle as described in the committee's charge.

2. Any procedure developed to obtain redress for any member who may or has suffered loss, financial or otherwise, as a result of positions taken on matters of an ethical nature must be such as to affect the IEEE resources only in a minimal fashion.

3. A feasibility study should be initiated to determine if suitable insurance coverage could be obtained so that members could be partially recompensated if it were showed

that they suffered loss of income as a result of actions taken on matters of ethical principle. Such a study would have to await Board of Directors approval of procedures to be employed in such cases.

4. Upon evaluation of the current thrust of the employment practices committee of the United States Activites Committee on matters of ethical principle and adherence of the members to or their employers to the employment guidelines in the IEEE code of ethics, the preliminary plan being developed was deemed to be responsive in dealing effectively with the charge to this committee as well as with other matters. The USAC committee on employment practices should be encouraged to continue to develop their plan and to submit it to the Executive Committee and to the Board of Directors.

Director Saunders presented five recommendations to the Executive Committee, and on December 4, the Executive Committee adopted two of them:

1. That USAC be instructed to report, for the action by the Executive Committee and Board, procedures to handle reports of non-adherence to the Employment Guidelines, and the Code of Ethics by March 15, 1975.

2. That the _ad hoc_ committee be discharged.

The other three recommendations were also approved by the Executive Committee for recommendation to the Board of Directors, and were approved unanimously by the full Board at its December 5-6, 1974 meeting. They are:

1. Executive Committee is empowered by the Board of Directors to enter an _amicus curiae_ brief in any court in the USA or in cooperation with cognizant national societies in other countries where a member of the profession is involved as a consequence of his taking a position on a matter of ethical principle.

2. The Executive Committee is empowered to publicize actions described in recommendation 1 in any fashion deemed suitable and appropriate.

3. It is Institute policy that the IEEE will not, as to disputed acts, intervene or take an adversary position on behalf or against any member involved in a matter of ethical principle.

Following the December, 1974 Board meeting, Frank Cummings, the legal counsel for USAC, worked very closely with Joel Snyder to prepare a draft of the <u>amicus curiae</u> brief. It was during this time that the code of ethics adopted in 1912 by the American Institute of Electrical Engineers (one of the original societies of the IEEE) was remembered, rediscovered, and incorporated into the brief as a basis for establishing proper ethical behavior for employed engineers. (It is interesting to note that at the December, 1974 Board meeting the Board also approved a new IEEE code of ethics. This new code of ethics was approved without knowledge of or reference to the 1912 code of ethics of AIEE).

On January 9, 1975, Jill Cummins of Gall, Lane, and Powell, Washington, D.C. appeared before presiding Judge George W. Phillips, Jr. of the Superior Court of California, County of Alameda, and petitioned for an order granting leave for IEEE to file an <u>amicus curiae</u> brief. The judge signed an order granting the petition, and the brief was filed. A copy of the brief is included in the appendix.

From a legal point of view the most significant aspect of the brief is contained in its conclusion.

> Based on the foregoing, we submit and we urge this court to acknowledge that an engineer has an overriding obligation to protect the public.
>
> Specifically we urge this court:

1. to rule that evidence of professional ethics is relevant, material and admissible in this case; and

2. to rule, as to any motions for judgment or any jury instructions, that an engineer is obligated to protect the public safety, that an engineer's contract of employment includes as a matter of law, and implied term that such engineer will protect the public safety, and that a discharge of an engineer solely for unsubstantial part because he acted to protect the public safety constitutes a breach of such implied term.

Many people, both disinterested and involved in the case, believe that it is unfortunate that the engineers' lawsuit was settled out of court, since that action forestalled the possibility of getting a judicial ruling on the points of law as suggested in the conclusion of the IEEE amicus curiae brief. Nonetheless, on January 29, 1975, the engineers' suit against BART was settled out of court for $25,000 less 40% lawyers fees for each of the three plaintiffs.

Frank Cummins comments on the settlement in an internal memorandum,

> I spoke today with Quint, attorney for the engineers. Just before trial BART evidently realized they had a loser and offered a very substantial settlement which the plaintiffs accepted. The amount of the settlement is being kept secret. I would guess that BART is a little embarrassed by it.
>
> Of course this means that the case itself will never come to trial and there will be no ruling by the court on the ethical and legal questions involved. On the other hand, I believe, as Quint believes, that the filing of our brief was a material factor in the outcome.

Part IV
Summary

Chapter 14

Conclusion

In the first chapter of this book, we raised a number of fundamental questions about the structure and functioning of large technologically-oriented organizations in modern society, as well as about the behavior of professional employees of these organizations. We suggested that an understanding of these questions was a necessary theoretical precursor of analysis of the series of events which has come to be called the BART incident. The rest of the book then chronicled, from several points of view, the events of the incident itself.

For these thirteen chapters, then, you have been asked to follow the genesis, the development, the climax and the denouement of what was an irritating minor crisis to the BART organization and a major--almost tragic--incident in the lives of the three BART engineers. Three of the chapters treated the incident from the perspective of the engineers and three from that of the organizational managers who finally fired them. The role of the Board of Directors was portrayed in three more chapters and the interests of the professional--technical societies in a final three. Although none of these groups was monolithic, some--BART management, for example--were more homogeneous in attitude than others. Therefore, some of the chapters, offering the perspectives of one or another of the groups, may give an impression as much of advocacy as of analysis. In all the chapters, however, we were trying to answer the same two questions: How did a particular series of events appear to a particular group of actors in the incident? Why were the events seen in this way?

Our original resolve was to go no further than this. We would tie up the loose ends in a brief epilog--a "where are they now" end note--but we would leave it up to the reader to tease the critical ethical and pragmatic issues out of the complex narrative, to make his own judgments about right and wrong and to distribute praise and blame as he saw fit.

But when we finally arrived at this point, it became clear to us that this was not enough. More was required of responsible students and analysts of the BART affair. This *more*, however, was not what one or another of the actors in the drama might have wished for, that is, an indictment of some and a vindication of others. Not that this would not have been easy. It would have required only a mildly selective arrangement of verifiable information to demonstrate that the Board of Directors was a hard-working, well-informed, disinterested band of dedicated citizens, committed totally to the successful completion of a project of great public interest--or that they were a group of relatively ignorant functionaries serving narrow and parochial interests, who permitted themselves, as a group and as individuals, to be gulled into irresponsible indolence or reckless action by unscrupulous management and self-serving employees. It would have been equally easy--and convincing--to demonstrate that the three engineers were martyrs to the cause of the public safety: capable and scrupulous engineers driven to take the actions they did by devotion to the noblest ideals of their profession and hounded from their jobs by a callous and cynical management. Or, on the contrary, to show that the engineers were limited and narrow specialists, goaded by a combination of technical arrogance, overweening ambition, and naivete bordering on obtuseness to engage in acts of treachery which threatened to destroy the acknowledged and applauded esprit of the BART organization. Similarly, we could have developed the evidence to show that management--harassed by a shallow and fickle public and a vicious and spiteful press--listened with remarkable forbearance to the complaints of the engineers, patiently took all appropriate actions, and dismissed them only in the face of their blatant insubordination, lack of candor, and their deliberate and malicious effort to rend the fabric of employee morale. Or yet, to demonstrate that management was hierarchical and authoritarian, technically inept and organizationally unresponsive, interested solely in protecting its past errors with present repression and capable of carrying its vendetta to the point of blacklisting the engineers and cruelly depriving them of other employment. Or, finally, to prove that the professional-technical society, the last, best hope of the ethical engineer, was willing to invest time, energy and money in the struggle to defend his rights and, by so doing, to maintain and protect the highest standards of the profession; or that the societies were no more and no less than their spokesmen--not infrequently self-selected--who, for a variety of irrelevant and probably ignoble reasons, jumped on the chic bandwagon of professional ethics and got in their licks against an organization struggling for its life against a veritable sea of troubles.

In the final analysis, none of these interpretations would do. All of them, after all, rest on judgments about people's characters. To justify these interpretations, we would have had to conclude that certain actors in the BART drama were malicious, arrogant, indolent, spiteful, prideful, and envious. To have decided that the roots of the problem lay in the natural and universal infirmities of human nature would have been tantamount to offering the reader and student of the case little more than an annotated catalogue--with the possible exception of lechery-- of the Seven Deadly Sins. While this might be appropriate for a theological tract, it offered us little opportunity to suggest constructive strategies for avoiding similar unfortunate conflicts and confrontations in the future.

When we began the BART study, it seemed like a reasonably clear case of conflict over a matter of professional ethics, rooted in the engineers' dedication to the ideal of safeguarding the public health and safety. After months of studying thousands of pages of public and private records and documents: newspapers and magazine articles, legal depositions, technical reports, letters, memoranda, personal interviews, Board minutes, Public Utility Commission hearings and a host of others, we can conclude with some assurance that the problems which inevitably precipitated the conflict were centrally neither those of professional ethics nor of concern for the public safety, and almost certainly were not inevitable manifestations of flawed human character. Simply put, most of those involved in the BART case appear to have been decent people who wanted to do their best in successfully completing a costly, complex and difficult project. How then, and why, did the problem arise? Although the series of actions which culminated in the dismissal of the engineers began in the Fall of 1971, the origins of the incident go farther back. Farther back than May, 1971, when Max Blankenzee came to work for BART, farther back than 1969, when Robert Bruder joined BART, or 1966, when Holger Hjortsvang left his job at Nuclear Research Instruments to throw in his lot with the exciting and glamorous new transit system. The origins are not to be found in any one set of organizational policies and procedures, nor in the administrative style or professional attitude of any group of managers or engineers. To a marked degree, the origins lie in the history and the environment of the whole Bay Area.

The public of the Bay Area never overwhelmingly supported the idea of BART. Many citizens of Marin, San Mateo, Contra Costa and Alameda Counties saw the system, with some justification, as an enormously expensive attempt to maintain the economic health of the city of San Francisco. Back in the fifties and early sixties they were content to rely on the old Key system of trains, on commuter buses and on their own private automobiles to travel from the suburbs to the city for work, shopping and recreation. Many citizens of these four counties resented having to pick up a major share of the costs of a system which would yield only marginal benefits to them, if, indeed, the benefits in the final analysis outweighed at all the costs in terms of tax dollars, loss of property and environmental injury. It was precisely because of this attitude that the supporters of BART effected a change in legislation governing public works bond issues, reducing the required minimum level of popular support in the referendum from two-thirds to 60% and providing that even this level not necessarily be met by the voters of each participating county, but be arrived at as a total of all Area votes. As expected, the heavier support of San Francisco County pulled the total vote slightly above the required minimum. Even with this change in rules, BART lost San Mateo and Marin Counties to the system before the vote when their respective County Supervisors decided against participation.

At the very start, then, BART was faced with the enormous obstacle of a large and vocal group of opponents among those whose taxes were paying for the system. As a result, it seemed obvious to the infant system's management that the great initial push had to be made in the selling of the system. If the new line was going to be perceived as just another commuter railroad, it was doomed. It was necessary, therefore, to build a new and exciting image. Much of the image-building remained on the verbal level only. The non-underground sections of the line, for example, were not called the elevated portion of the system. Instead, they were referred to as the "supported duorail sections," terminology which conjured up associations with such glamorous technology as suspended monorails.

If the selling of the system had not gone beyond mere verbal cosmetology, probably no real harm would have been done. But it did. In order to impress the public with BART's novelty and uniqueness, the Board of Directors and management specified a completely automatic train control (ATC) system, a glamorous space-age but needlessly complicated article of technology. Their specification of this system should not, however, be a basis for indicting BART's directors and managers. It probably means simply that they shrewdly and accurately diagnosed the public's appetite for the excitement of novel gadgetry, and in feeding it, succeeded in making the whole BART system reasonably palatable--for a time, at least.

A number of observers of the BART incident--mostly engineers by profession--have remarked with some surprise on the Directors' choice in 1963 of B. R. Stokes, a former newspaperman and BART's first public relations director, as General Manager of the system. To any perceptive analyst of BART's ambiance and clientele, however, this choice should have come as no shock. A key characteristic of the whole BART phenomenon was that it was, to a considerable degree, an essay in public relations and public persuasion. Why not, then, choose as its chief executive officer one who had made a profession of informing and persuading the public?

Given all these circumstances, it would have been extraordinary if the public relations gilding had not diffused below the surface of the project and dictated some of the system's technical and design properties. And so it did. The appearance of some of the stations, the complexity of the ATC and the design of the cars--first the planning of cumbersome removable pods for the train operators and then the development of two types of cars--A and B--which needlessly complicated the switching, lengthening and shortening of the trains--all seem to have been prescribed by the image, rather than the technical demands of BART.

This need constantly to satisfy the public's taste for glamour and to explain, justify and apologize for every planning decision was a direct result of the demands of the Bay Area public--and, very likely, any 20th Century American public--that new transportation technology be flashy and exciting--but also fast, safe, reliable and cheap. This was the challenge faced by BART's Board and Management: to try to deliver to the Bay Area at low cost a beautiful and efficient public facility which always ran through or next to someone else's property--and to get it done tomorrow morning. When it didn't all happen this way, large segments of the public and the press came down like a ton of bricks on the heads of the system's managers and directors.

The first pre-built ingredient of the BART crisis, then, was the public environment in which the system was born and grew. A second ingredient was the organizational structure and procedures which were set up to achieve BART's goals.

A basic problem was that the BART organization lacked a legal form capable of making the system financially self-sufficient yet directly responsible to its source of funds, the public. Without adequate independent taxing (or bonding) authority, BART became a beggar, petitioning agencies at the local, state or federal level for funds or for fund-raising authority. This basic weakness had profound effects on the total organization and its relationships with its publics.

A second difficulty was that the members of the Board of Directors were politically appointed, either by a County Board of Supervisors or a Conference of Mayors. Despite the most honest and devoted allegiance to the public interest which individual Board members might have had, the combination of political appointment and BART's financial dependency forced the directors to be parochial advocates rather than shapers of a true policy consensus. Because of the narrowness with which the directors were forced to represent the public, the entire Board never really faced up to the task of establishing a set of organizational controls which would, on a continuing basis, monitor operations and keep policy decisions and management actions within economic, temporal and technological limits.

As a consequence, management was put in the position of working with a politically fragmented and, therefore, relatively ineffective Board. Operating without the restraining influence of an active Board, management simply moved by default into the power vacuum in order to make the decisions which had to be made.

Finally, BART's enabling legislation simply placed too much power and authority in the hands of its General Manager. He was granted statutory authority to do too much by fiat, without the prior assent or approval of the Board. It is not possible that a man like Bill Stokes--universally recognized to have possessed the intelligence, the initiative, the drive and the dedication to make BART work--would have failed to use all the power and all the authority available to him in trying to achieve the organization's goals. The whole history of Board assent to management initiatives discloses, therefore, not a band of power-voracious managers nor an indifferent, slothful Board, but rather a set of organizational structures and procedures which were simply not adequately adapted to the demands made upon them.

Finally, the structure of the relationship between BART's managers and professional employees and the PBTB engineering consortium was almost preordained to give rise to situations like that of the three engineers. The principle underlying the BART-PBTB relationship seems both sound and reasonable. BART required the efforts of a multitude of engineers and technicians during the design and construction phases of the operation. A much smaller professional staff, however would be needed after the start of revenue service for operation and maintenance of the system. Even today, former top managers of BART insist that the management principle adopted was the best possible one, and one which they would use if they were again faced with the same task. And to the degree that it avoided the personnel problems and the major employment dislocation which would have occurred if BART had hired--and then had to fire--a large number of engineers, the principle was sound. The only problem was that the working relationships created by the arrangement with PBTB were almost designed to cause real job frustration among BART's own engineers.

BART's engineers had few directly assigned responsibilities. Almost all of them testify to the high degree of discretion each of them had in going about his daily work. Simply put, the BART engineers had no really operational responsibilities. Rather, in a loose fashion, they were responsible for supervising PBTB's execution of its contractual obligations--but in a very general way. They were used, in short, more as management spies than as parts of an integrated and functioning technical team. BART's management suggests, however, that this view shows the degree to which some engineers misinterpreted their job responsibilities, prime among which, says management, was to use the construction phase as a learning situation preparing them for operating responsibilities later on. In this role, whether construed as spies or learners, they had every opportunity to observe the inevitable problems which arise in any new engineering endeavor, but had little or no chance to do anything about them. Their observations were reported to their supervisors and to conferences of BART and PBTB engineers--but that is where the BART employees' responsibilities ceased. Whether or not the issues raised by the three engineers turn out in retrospect to have been major or minor, central or peripheral to the issues of safety and reliability, does not really matter. The fact is that any comments offered by the BART engineers were perceived by them to go unheard and unattended to, despite the most vehement denials by BART's supervisory staff. The BART engineers, in other words, felt professionally frustrated and underutilized because of the role into which BART-PBTB contractual structure cast them.

Coupled with this failure of the organization to involve its professional employees in meaningful technical and decision-making roles was the sense of impending disaster which the PBTB connection cast over them. Despite all assurances to the contrary, the BART engineers believed that, when their design and construction obligations were satisfied, the PBTB and Westinghouse engineers would simply pull up stakes and go on to another job, leaving the job of operating a new and extremely complex train control system in their hands. The BART engineers simply felt that they did not know enough about the basic design of the system to accept this responsibility. They did not want to have dumped on them a railroad they did not know how to run.

The three engineers were, of course, all individuals, with unique personal and professional histories, characters, temperaments and ethical standards. Motivations to action are rarely pure and unitary, and when the engineers maintain that the claim of the ideal was their sole mover they are probably as selective in their analysis as management, which sees simple self-aggrandizement as the engineers' spur to action. It may have taken the catalytic presence of the young and forceful Max Blankenzee to focus and set into motion the normal--and complex--mix of attitudes and motives held by Hjortsvang and Bruder--ambition, genuine public interest, self-interest, nagging frustration.

The fact that the three did what they did in the way they did may underscore the one characteristic held in common by many technical professionals: their tendency to ignore the role of non-technical influences upon the decision-making process, a tendency which translates into a high degree of political naivete. Finally, the ingredient in the situation which permitted the whole incident to get out of hand was once again prompted by the basic structure of the BART Board and organization. This was the bounding onto stage of the newly appointed Board member, Dan Helix. Because of the political nature of the BART Board appointments and the differential attention given to the various communities to be served by BART, Helix was a man with a mission--perhaps the only Board member actually to have campaigned for his seat. Given his grievances toward BART as a councilman in outlying Concord and his own naive enthusiasm for finding a club with which to beat on a management he was convinced was not committed to fair treatment for all BART's constituents, Helix was willing--even eager--to take up the case of the engineers and run with it when two more experienced Board members--Blake and Bianco--who were probably sympathetic, were well warned by their political sensitivities to stay away from it.

Finally, the professional-technical societies were unsuccessful either in clarifying the issues or in proposing solutions because they failed to see the basic problem. There is no question that many of the society members who were active in the case were motivated by high ethical principles. The roles they played suggest, however, that they saw the problems in isolated and simplistic fashion. On the one hand, some of them saw an insistent concern for the public safety as the overriding ethical imperative for the employed engineer. By assuming the disinterestedness and integrity of the whistle-blowing engineer, these advocates are inevitably led to the conclusion that management and/or the Board of Directors were guilty of callous disregard for the public safety and are, therefore, morally culpable. On the other side, certain members of the professional and technical societies perceived the impugning of the professional integrity of one group of engineers--management--by another--the three engineers and their supporters in the CSPE--as being a serious breach of professional ethics. In both cases, it appears the concern focusses on the specific actions of individuals, considered quite apart from the total organizational context.

Viewed in isolation, both actions can, of course, be construed as violations of professional ethical norms. The key issue, never articulated by the societies, is that the ethical codes of the profession, as they are written, are simply incapable of dealing with most conflicts which arise within the structure of large and complex organizations employing professionals at a variety of levels. At no time, for example, did the spokesmen for the societies distinguish between engineers in the organization practicing at a purely technical level and those whose responsibilities were primarily administrative and supervisory. Nor was there any sensitivity to the possibility that the structure of the organization, the allocation of power, authority and responsibility and the total ambiance of the BART project forced the actors into various roles which made conflict almost inevitable.

As a consequence, any ethical prescriptions which emerge from the societies as a result of this and comparable incidents involving professional behavior are likely to be incontestable generalizations which still fail to give specific and reasonable guidance to participants in real life organizational and ethical conflicts.

In light of this conclusion, is it possible to offer any prescriptions or to articulate any standards of behavior for organizations or individuals which would make the development of similar incidents less likely? To the degree that individual behavior is almost inevitably linked to self-interest, and that a variety of incompatible self-interests are likely to eventuate in conflict, we cannot offer an optimistic answer to this question. For example, although we find no villains (and incidentally, few heroes) in the BART story, and although the people involved seem generally to be decent people trying hard to do a good and responsible job, almost all were motivated in their actions by at least a modicum of self-interest. One of the engineers aspired to head his own engineering group--and the others appear to have been at least normally ambitious professionals. Board members' personal interests were inextricably entangled with the interests of the political regions they represented and some, like Helix, had personal political ambitions which were not sharply separable from his role as a Board member. Management had for so long been criticized by the press and by anti-BART forces in the community that their self-interest demanded a posture which protected and defended the wisdom of their prior decisions: setting up the relationship with PBTB, contracting the ATC System to Westinghouse and the car construction to Rohr, and specifying a fully automatic train control system in the first place. Some of the members of the CSPE most vocal in their criticism of BART and in defense of the three engineers had opposed BART in the past for a variety of personal and professional reasons, and it is unreasonable to assume they would not see this case as justification of their earlier opposition, a view making them gravitate naturally to the side of the engineers. The IEEE was at the point organizationally when it wished to grow beyond its image as a technical society and move into broader, more professional areas. What better basis for doing this than an issue involving sacrosanct and incontrovertible matters like ethical behavior and the public safety? State legislators who, as members of the Senate Public Utilities and Corporations Committee, held hearings on the issues of safety and BART management, also had some personal interest in becoming involved. As purely motivated as they may have been to protect the public's safety and its financial investment, they were also elected officials, and no politician forswears the opportunity to go on record, and to be portrayed in the press, as a defender of public safety and proponent of public frugality.

Given these personal investments by the various actors, there can obviously be offered no prescription guaranteed to avoid organizational conflict. To the degree, however, that these personal investments were made in a situation with particular predisposing structural characteristics, it is possible to suggest certain organizational strategies designed to minimize the liklihood of their creating overt conflicts.

First, the allocation of policy- and decision-making power and authority should be clearly and unambiguously indicated. In BART, this power was inappropriately divided between the Board and the General Manager, and so the Board--established in such a way as to emphasize the political and parochial demands on all its members--was never able to function in a true policy-making role, and was incapable of controlling or containing the issue of the three engineers once it erupted. As a matter of fact, as exemplified in the behavior of Dan Helix, the unclear policy-making activities of the Board clearly aggravated and exacerbated an already difficult situation.

Second, management shares with the three engineers responsibility for the political naivete which permitted them to carry their grievance as far as they did. It is clear that the engineers took a narrow and technical view of the issues which disturbed them, and failed to place them in the context of the whole BART development. At the same time, management fostered this naivete by failing adequately to sensitize its professional employees to the political and economic climate surrounding and influencing the activities of the organization. In dealing with professional employees, it appears that management would have greater success at preventing conflicts of this sort if it avoided emphasis on narrow specialization of roles, but rather attempted to orient employees concerning the broad context in which the organization operates, and at least informed them of, if it did not share with them, some of the general decision-making processes of the organization.

Third, although the contracting out of major design and construction responsibilities to a consultant like PBTB does make sense, it appears that BART failed to recruit its own supervisory staff in such a way that it could effectively monitor the work of PBTB and its subcontractors. The over heavy reliance of BART managers on PBTB's conclusions and recommendations shook the confidence of the BART engineers in the technical competency of their own supervisory personnel. Although it is not clear whether the fault lay chiefly in the terms of the contract between BART and PBTB or just in BART's personnel recruiting style and schedule, any organization charged with high-level technical supervision should be assured that the people employed to supervise are really competent to do so, and that the procedures of the organization and the allocation of authority permit them to carry out their jobs.

Fourth, organizations should view their purposes and conduct themselves in such a way as to avoid incompatibility between real goals and stated goals. Specifically, it appears that the real goal of the BART system was the renewal and invigoration of San Francisco as a commercial center. As such, it was strongly supported by the financial and business interests of the city. Because of strong opposition to BART in the four surrounding counties, however, it became necessary to sell the system on the basis of other appeals. Hence, the development of a glamorous, space-age car and control system specifications which needlessly complicated both the system's technical development and public posture. Much more conventional technology could have been contracted for, designed and built at a lower cost and in a shorter time--but such a system could probably not have received a 60% vote in the bond referendum. Since it was the complexities of the novel control system and the elaborate and impressive design specifications of some of the stations which were, to some degree, responsible for the time delays and cost overruns which underlay much of the behavior of the engineers, management and the Board, we see a case of the public relations tail wagging the design and engineering dog--and being fundamentally responsible for many of the woes of the directors, managers and professional employees of BART.

Finally, what was ostensibly a simple case of public safety and engineering ethics can be viewed as not really involving safety or ethics to any marked degree. Rather, it is a case of a number of actors being placed in roles by the structure of a large organization and its political, social and economic environment which, functioning together, almost inevitably created conficts. These structurally engendered conflicts then almost seemed to cast about for a proper raison d'etre--and the issue of public safety filled the bill.

In a society which will apparently see more quasi-private, quasi- public, "third sector" organizations emerging to respond to public policy demands, the errors of the BART experience can serve as a valuable object lesson for planners of the future.

Appendixes

Appendix A

BART Board of Directors

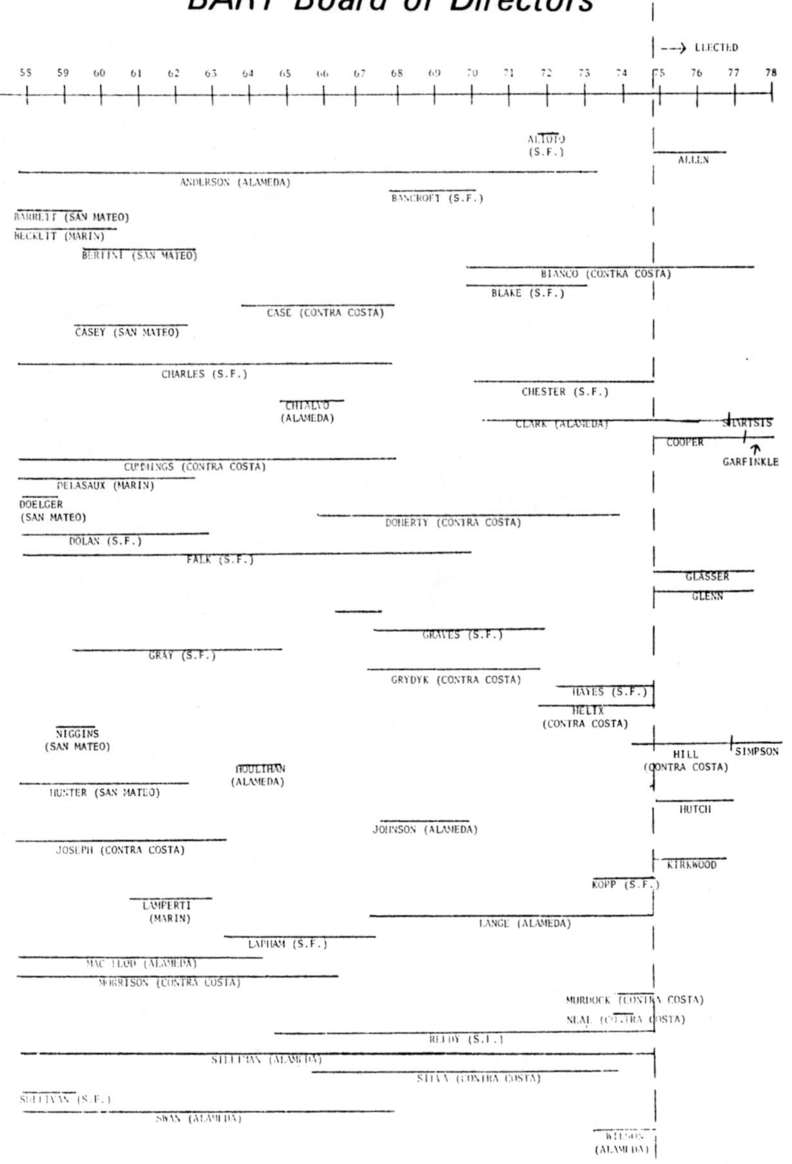

Appendix B
Partial Table of Organization

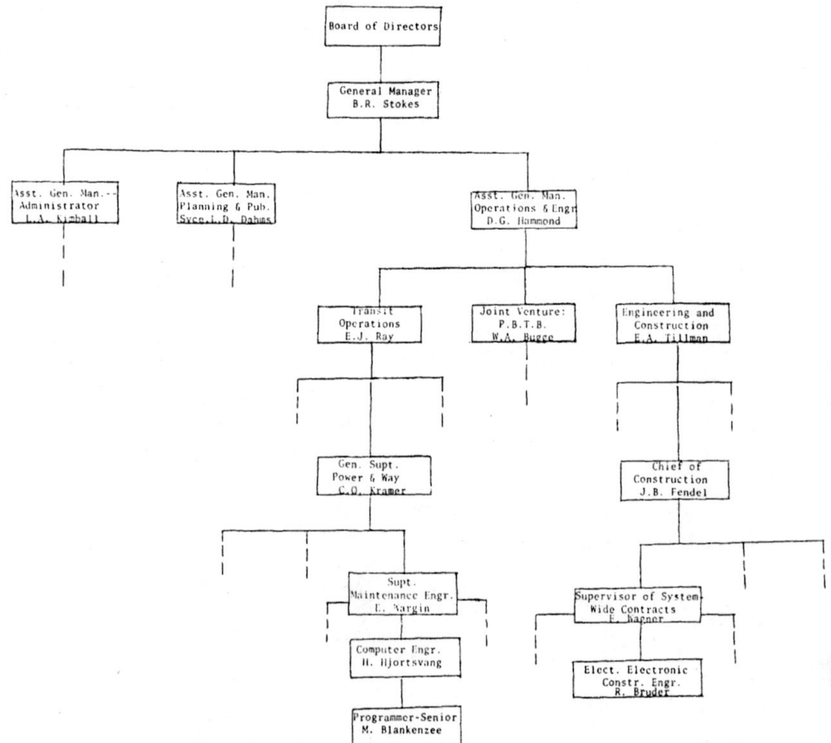

Appendix C

List of Persons Mentioned in Text

Alphabetical List

Names appearing in the BART monograph

(Note: Each person is identified by the position he held during the BART incident.)

Deane N. Aboudara: Maintenance Dept., BARTD. He supervised John Andrews and Charles O. Kramer.

Frank N. Alioto: Director from San Francisco County, on the BARTD Board of Directors from Nov. 30, 1971 until Feb. 29, 1972.

Alfred E. Alquist: Senator, Thirteenth District, California State Legislature. He was the chairman of the California State Senate Committee on Public Utilities and Corporations which investigated BARTD after the firing of the three engineers.

Arnold C. Anderson: Director from Alameda County, on the BARTD Board of Directors from Nov. 14, 1957 until April 10, 1973.

Roy W. Anderson: State director, president and vice president during 1964-72 of the California Society of Professional Engineers (CSPE) Diablo chapter; Chairman of CSPE's Transportation Safety Committee; Co-author of "The BART Inquiry".

John Andrews: Senior Electronics Engineer, BARTD. He was Holger Hjortsvang's first supervisor at BART.

Dr. Bruce A. Anthony: Executive secretary to the California Society of Professional Engineers Board of Directors.

John R. Asmus: Engineering Manager, Parsons, Brinckerhoff, Tudor & Bechtel (PBTB).

Lowell G. Banks: BART employee.

Frank Barns: Member, IEEE Ethical Practices Committee in 1974

Peter H. Behr: Senator, California State Legislature; Member, California State Senate Committee on Public Utilities and Corporations in 1974.

Nello J. Bianco: Director from Contra Costa County, on the BARTD Board of Directors from Oct. 22, 1969 until Nov. 24, 1978.

Irwin W. Black: Roy Anderson's supervisor.

William C. Blake: Directors from San Francisco County, on BARTD Board of Directors from Nov. 24, 1969 until Jan. 8, 1973.

Max Blankenzee: Senior programmer analyst, BARTD; hired 1971, fired March 2, 1972.

Mark Bowers: Director of Employee Relations, BARTD

Dr. William M. Brobeck: President, William M. Brobeck Assoc.; Member, Blue Ribbon Committee of the California State Senate Public Utilities and Corporations Committee.

Robert Bruder: Electrical/Electronic construction engineer, BARTD; hired 1969; fired March 3, 1972.

William A. Bugge: Project director for PBTB for the BART project.

Edward A. Burfine: Beckers, Burfine & Associates, Palo Alto, CA; author of "The Burfine Report" critical of BARTD.

Burnham: Chairman of the Board, Westinghouse Electric Corp.

Jay S. Burns: Engineering Dept., BARTD.

William H. Chester: Vice President and Director from San Francisco County, on BARTD Board of Directors from Jan. 23, 1970 until Nov. 29, 1974.

Richard O. Clark: Director from Alameda County, on BARTD Board of Directors from March 4, 1970 until Nov. 26, 1976.

Jack Cornwell: Disney Industries; David Snyder's supervisor.

Frank and Jill Cummings: IEEE attorneys, prepared IEEE amicus curiae brief.

H. L. (Jack) Cummings: Director from Martinez, CA, on BARTD Board of Directors from Sept. 10, 1957 until Oct. 22, 1969.

A. A. Daush: Vice President South, California Society of Professional Engineers.

James P. Doherty: Director from Contra Costa County, on BARTD Board of Directors from Sept. 30, 1965 until Sept. 29, 1973.

Jerome C. Dougherty: Attorney for BARTD

C. H. Engle: BART employee.

Adrien J. Falk: Director and President (in 1968) on BARTD Board of Directors from Oct. 28, 1957 until Nov. 24, 1969.

John B. Fendel: Chief of Construction, BARTD; Frank Wagner's supervisor.

R. A. Fickes: BART computer staff.

Donald Fink: IEEE Executive Director

Dr. Thomas L. Follett: Computer Science Dept., University of California, Berkeley; reviewed BART's computer system.

Gordon Friedlander: Senior Staff Writer, IEEE *Spectrum*; wrote a series of articles on BARTD.

Harold Goldberg: President, Data Precision Corp., Wakefield, Mass.; Member, IEEE.

Howard P. Grant: Past president, California Society of Professional Engineers Golden Gate Chapter; Member, CSPE Ad Hoc Committee on Ethical Practices.

John J. Guarrera: President, IEEE (when the decision was made to file the amicus curiae brief in the engineers' court case)

John Haff: Member, California Society of Professional Engineers Ad Hoc Committee to act on the BART matter.

David Hammond: Asst. General Manager, Operations and Engineering, BARTD; resigned March 1, 1973.

Daniel Helix: Director from Contra Costa County, on BARTD Board of Directors from Oct. 28, 1971 until Nov. 29, 1974.

S. W. (Cy) Herwald: Vice President, Westinghouse Electric Corp.

Holger Hjortsvang: Train Control engineer, BARTD; hired 1966; fired March 2, 1972.

Bill Holden: Member, California Society of Professional Engineers Ad Hoc Committee to confer with the three BART engineers.

B. J. Honholt: Treasurer, California Society of Professional Engineers.

Ken M. Hoover: Director of Engineering, PBTB; Dean Aboudara's supervisor.

George Humphrey: Member, California Society of Professional Engineers Diablo Chapter.

E. C. (Ernie) James: Vice President Central, California Society of Professional Engineers.

Dr. Woodrow E. Johnson: Vice President & General Manager, Transportation Division, Westinghouse Electric Corp.

William F. Jones: President, California Society of Professional Engineers, 1971-72.

John Kane: Editor, NSPE's *Professional Engineer*.

Charles O. Kramer: General Superintendent, Power and Way, BARTD; Ed Wargin's supervisor.

Robert Kuntz: President, California Society of Professional Engineers, 1972-1973.

James F. Lackey, Jr.: President, Diablo Chapter, California Society of Professional Engineers.

Harry N. Lalor: National Director, California Society of Professional Engineers, 1972; Member, CSPE Ad Hoc Committee to act on BART matter.

Harry R. Lange: Director from Alameda County, on BARTD Board of Directors from Sept. 5, 1967 until Nov. 29, 1974.

John R. Lavinder: Lafayette, CA industrialist who ran for the BARTD Board of Directors at the same time as Roy Anderson.

D. Y. Lee: BART employee.

Dean S. Lesher: Publisher, *Contra Costa Times*.

Luther Lincoln: Former California State Assembly speaker from Concord, CA who ran for the BARTD Board at the same time as Roy Anderson.

Dr. Clarence A. Lovell: Consultant, Fairfax, VA; Chairman, Blue Ribbon Committee of the California State Senate Committee on Public Utilities and Corporations

Milton Lunch: NSPE attorney.

Charles E. Luckhardt, Jr.: Attorney retained by CSPE to represent them in filing an amicus curiae brief in the case of the three engineers.

Ashley D. Martin: President Chapter 32, California Society of Professional Engineers.

J. Eugene McAteer: Senator, California State Legislature.

David Miller: Vice President, Daniel, Mann, Johnson and Mendenhall, Baltimore, Maryland; interviewed Holger Hjortsvang for a job after he was fired from BART.

Howard N. Miller: Engineering Manager of the Transportation Division, Westinghouse Electric Corp.

James R. Mills: Senator, California State Senate; Member, California State Senate Committee on Public Utilities and Corporations.

Dana Murdock: Director, on BARTD Board of Directors from Oct. 22, 1973 until Nov. 29, 1974.

John A. Nejedly: Senator, Seventh District, California State Senate; Ex-Officio Member, Senate Public Utilities and Corporations Committee.

Robert Nichols: Chairman, NSPE Professional Engineers in Private Practice (PEPP).

Dr. Bernard M. Oliver: Vice President, Research & Development, Hewlett Packard Co.; Member, Blue Ribbon Committee of the California State Senate Committee on Public Utilities and Corporations.

Phillip O. Orsmbee: Director of Public Relations, BARTD.

Gilbert Ortiz: BART employee, called a "union organizer"

Elliott Pearl: Legal counsel for CSPE during BART controversy

Nicholas C. Petris: Senator, California State Senate; Ex-Officio Member, Senate Public Utilities and Corporations Committee

George W. Phillips, Jr.: Judge who heard the three engineers lawsuit in Superior Court, Alameda County, CA

John Pierce: First General Manager of BARTD, 1957-62.

A. Alan Post: Legislative Analyst appointed by the California State Senate Committee on Public Utilities and Corporations to investigate the safety and contract administration of BARTD in 1972.

Robert A. Profet: Advanced Systems engineer on leave from McDonnell-Douglas Corp.; headed an internal panel to review the automatic control system;

fired from the BARTD Board of Directors after his critical report.

Walter P. Quentin: Supervising Control Engineer, PBTB
Matthew F. Quint: Attorney for Hjortsvang, Bruder, and Blankenzee.

C. W. Rae: BART employee.
Edward J. Ray: Director of Operations, BARTD; C. O. Kramer's supervisor.
William M. Reedy: Director from San Francisco County, on BARTD Board of Directors from Oct. 26, 1964 until Nov. 29, 1974; Chairman of the Engineering Committee of the Board, 1972.
Louis W. Riggs: President, Tudor Engineering; on the Board of Control of PBTB for the BART project.
Jim Rippol: Secretary, Diablo Chapter, California Society of Professional Engineers.
Justin Roberts: Reporter, Contra Costa Times.

Mike Sassano: Member, IEEE.
Robert M. Saunders: School of Engineering, University of California--Irvine; Director, IEEE.
B. A. (Billy) Schmidt: Vice President North, California Society of Professional Engineers; Member, CSPE Ad Hoc Committee to act on the BART matter.
Leendert Schoneville: First Vice President, California Society of Professional Engineers; President, CSPE (two terms following Robert Kuntz).
Ralph Sheldon: Train Control Engineer, Washington Metropolitan Transit Authorities.
James F. Shivler: President, NSPE.
George M. Silliman: Director from Alameda County, on BARTD Board of Directors from Oct. 28, 1957 until Nov. 29, 1974; President, 1972.
Joseph S. Silva: Director from Contra Costa County, on BARTD Board of Directors from Oct. 22, 1965 until Oct. 21, 1973.
Traver J. Smith: Chairman, California Society of Professional Engineers Ad Hoc Committee on Ethical Practices.
David Snyder: Walt Disney Productions, Burbank, CA; interviewed Blankenzee for a job after he was fired from BART.
Joel B. Snyder: Member, IEEE.
Arthur Stern: President Elect, IEEE.
B. R. Stokes: General Manager, BARTD; resigned May 24, 1974.
Eugene F. Sullivan: President, Santa Clara Valley Chapter, Society of Professional Engineers.

William Tarman: Member, California Society of Professional Engineers; Member, CSPE Ad Hoc Committee to confer with the three engineers.
Lawrence A. Tate: IBM engineer who was fired on an ethics matter, brought Steve Unger's proposal to support the ethical engineer before the IEEE Employment Practices Committee.

Erland A. Tillman: Chief, Engineering and Construction, BARTD; John Fendel's supervisor.
Byron Tomlinson: Vice President State, California Society of Professional Engineers.
Ron Tsugita: President, East Bay Chapter, California Society of Professional Engineers.

Stephen H. Unger: Dept. of Electrical Engineering and Computer Science, Columbia University, New York City; Member, IEEE.

Gilbert A. Verdugo: President, Diablo Chapter, California Society of Professional Engineers, 1973.

Frank Wagner: Supervisor of systemwide contracts, construction section, engineering and construction dept., BARTD; Bruder's supervisor.
Edward Walker: President, Golden Gate Chapter, California Society of Professional Engineers, 1972-73.
James J. Walsh, Jr.: NSPE National Director, 1972-73.
Ed. F. Wargin: Superintendent of Maintenance Engineering, BARTD; Blankenzee's supervisor
Wasie: Bonneville Electrical Power Project, Portland, Oregon; interviewed Blankenzee for a job after he was fired from BART.
Dr. Willard H. Wattenburg: Consultant critical of BARTD.
J. S. Whitely: BART employee.
Harold E. (Hal) Williamson: Member, California Society of Professional Engineers Ad Hoc Committee on Ethical Practices, CSPE Ad Hoc Committee to act on the BART matter.
C. Glenn Wilson: Vice President South, California Society of Professional Engineers; headed CSPE Ad Hoc Committee to confer with the three BART engineers.
Ralph Woods: Washington Metropolitan Transit Authority; Ralph Sheldon's supervisor.
Jim Wright: Member, Western Council of Engineers.

Charles Yata: President, Chapter 12, California Society of Professional Engineers.
Leo Young: Chairman, U.S. Activities Board, IEEE, 1972.

Victor Zourides: Member, Long Island IEEE.

Appendix D

Code of Ethics

Code of Ethics

For Engineers

Preamble

The Engineer, to uphold and advance the honor and dignity of the engineering profession and in keeping with high standards of ethical conduct:

- *Will be honest and impartial, and will serve with devotion his employer, his clients, and the public;*
- *Will strive to increase the competence and prestige of the engineering profession;*
- *Will use his knowledge and skill for the advancement of human welfare.*

Section 1—The Engineer will be guided in all his professional relations by the highest standards of integrity, and will act in professional matters for each client or employer as a faithful agent or trustee.

a. He will be realistic and honest in all estimates, reports, statements, and testimony.

b. He will admit and accept his own errors when proven wrong and refrain from distorting or altering the facts in an attempt to justify his decision.

c. He will advise his client or employer when he believes a project will not be successful.

d. He will not accept outside employment to the detriment of his regular work or interest, or without the consent of his employer.

e. He will not attempt to attract an engineer from another employer by false or misleading pretenses.

f. He will not actively participate in strikes, picket lines, or other collective coercive action.

g. He will avoid any act tending to promote his own interest at the expense of the dignity and integrity of the profession.

Section 2—The Engineer will have proper regard for the safety, health, and welfare of the public in the performance of his professional duties. If his engineering judgment is overruled by nontechnical authority, he will clearly point out the consequences. He will notify the proper authority of any observed conditions which endanger public safety and health.

a. He will regard his duty to the public welfare as paramount.

b. He shall seek opportunities to be of constructive service in civic affairs and work for the advancement of the safety, health and well-being of his community.

c. He will not complete, sign, or seal plans and/or specifications that are not of a design safe to the public health and welfare and in conformity with accepted engineering standards. If the client or employer insists on such unprofessional conduct, he shall notify the proper authorities and withdraw from further service on the project.

Section 3—The Engineer will avoid all conduct or practice likely to discredit or unfavorably reflect upon the dignity or honor of the profession.

a. The Engineer shall not advertise his professional services but may utilize the following means of identification:

(1) Professional cards and listings in recognized and dignified publications, provided they are consistent in size and are in a section of the publication regularly devoted to such professional cards and listings. The information displayed must be restricted to firm name, address, telephone number, appropriate symbol, name of principal participants and the fields of practice in which the firm is qualified.

tion presented must be displayed in a dignified manner, restricted to firm name, address, telephone number, appropriate symbol, name of principal participants, the fields of practice in which the firm is qualified and factual descriptions of positions available, qualifications required and benefits available.

c. The Engineer may prepare articles for the lay or technical press which are factual, dignified and free from ostentations or laudatory implications. Such articles shall not imply other than his direct participation in the work described unless credit is given to others for their share of the work.

d. The Engineer may extend permission for his name to be used in commercial advertisements, such as may be published by manufacturers, contractors, material suppliers, etc., only by means of a modest dignified notation acknowledging his participation and the scope thereof in the project or product described. Such permission shall not include public endorsement of proprietary products.

e. The Engineer will not allow himself to be listed for employment using exaggerated statements of his qualifications.

Section 4—The Engineer will endeavor to extend public knowledge and appreciation of engineering and its achievements and to protect the engineering profession from misrepresentation and misunderstanding.

a. He shall not issue statements, criticisms, or arguments on matters connected with public policy which are inspired or paid for by private interests, unless he indicates on whose behalf he is making the statement.

Section 5—The Engineer will express an opinion of an engineering subject only when founded on adequate knowledge and honest conviction.

a. The Engineer will insist on the use of facts in reference to an engineering project in a group discussion, public forum or publication of articles.

Section 6—The Engineer will undertake engineering assignments for which he will be responsible only when qualified by training or experience, and he will engage, or advise engaging, experts and specialists whenever the client's or employer's interests are best served by such service.

Section 7—The Engineer will not disclose confidential information concerning the business affairs or technical processes of any present or former client or employer without his consent.

a. While in the employ of others, he will not enter promotional efforts or negotiations for work or make arrangements for other employment as a principal or to practice in connection with a specific project for which he has gained particular and specialized knowledge without the consent of all interested parties.

Section 8—The Engineer will endeavor to avoid a conflict of interest with his employer or client, but when unavoidable, the Engineer shall fully disclose the circumstances to his employer or client.

a. The Engineer will inform his client or employer of any business connections, interests, or circumstances which may be deemed as influencing his judgment or the quality of his services to his client or employer.

b. When in public service as a member, advisor, or employee of a governmental body or department, an Engineer shall not participate in considerations or actions with respect to services provided by him or his organization in private engineering practice.

(2) Signs on equipment, offices and at the site of projects for which he renders services, limited to firm name, address, telephone number and type of services, as appropriate.

(3) Brochures, business cards, letterheads and other factual representations of experience, facilities, personnel and capacity to render service, providing the same are not misleading relative to the extent of participation in the projects cited, and provided the same are not indiscriminately distributed.

(4) Listings in the classified section of telephone directories, limited to name, address, telephone number and specialties in which the firm is qualified.

b. The Engineer may advertise for recruitment of personnel in appropriate publications or by special distribution. The informa-

c. He will not accept remuneration from either an employee or employment agency for giving employment.

d. When hiring other engineers, he shall offer a salary according to the engineer's qualifications and the recognized standards in the particular geographical area.

e. If, in sales employ, he will not offer, or give engineering consultation, or designs, or advice other than specifically applying to the equipment being sold.

Section 10—The Engineer will not accept compensation, financial or otherwise, from more than one interested party for the same service, or for services pertaining to the same work, unless there is full disclosure to and consent of all interested parties.

a. He will not accept financial or other considerations, including free engineering designs, from material or equipment suppliers for specifying their product.

b. He will not accept commissions or allowances, directly or indirectly, from contractors or other parties dealing with his clients or employer in connection with work for which he is responsible.

Section 11—The Engineer will not compete unfairly with another engineer by attempting to obtain employment or advancement or professional engagements by competitive bidding, by taking advantage of a salaried position, by criticizing other engineers, or by other improper or questionable methods.

a. The Engineer will not attempt to supplant another engineer in a particular employment after becoming aware that definite steps have been taken toward the other's employment.

b. He will not pay, or offer to pay, either directly or indirectly, any commission, political contribution, or a gift, or other consideration in order to secure work, exclusive of securing salaried positions through employment agencies.

c. He shall not solicit or submit engineering proposals on the basis of competitive bidding. Competitive bidding for professional engineering services is defined as the formal or informal submission, or receipt, of verbal or written estimates of cost or proposals in terms of dollars, man days of work required, percentage of construction cost, or any other measure of compensation whereby the prospective client may compare engineering services on a price basis prior to the time that one engineer, or one engineering organization, has been selected for negotiations. The disclosure of recommended fee schedules prepared by various engineering societies is not considered to constitute competitive bidding. An Engineer requested to submit a fee proposal or bid prior to the selection of an engineer or firm subject to the negotiation of a satisfactory contract, shall attempt to have the procedure changed to conform to ethical practices, but if not successful he shall withdraw from consideration for the proposed work. These principles shall be applied by the Engineer in obtaining the services of other professionals.

d. An Engineer shall not request, propose, or accept a professional commission on a contingent basis under circumstances in which his professional judgment may be compromised, or when a contingency provision is used as a device for promoting or securing a professional commission.

e. While in a salaried position, he will accept part-time engineering work only at a salary or fee not less than that recognized as standard in the area.

f. An Engineer will not use equipment, supplies, laboratory, or office facilities of his employer to carry on outside private practice without consent.

Section 12—The Engineer will not attempt to injure, maliciously or falsely, directly or indirectly, the professional reputation,

c. An Engineer shall not solicit or accept an engineering contract from a governmental body on which a principal or officer of his organization serves as a member.

Section 9—The Engineer will uphold the principle of appropriate and adequate compensation for those engaged in engineering work.

a. He will not undertake or agree to perform any engineering service on a free basis, except for civic, charitable, religious, or eleemosynary nonprofit organizations when the professional services are advisory in nature.

b. He will not undertake work at a fee or salary below the accepted standards of the profession in the area.

prospects, practice or employment of another engineer, nor will he indiscriminately criticize another engineer's work. If he believes that another engineer is guilty of unethical or illegal practice, he shall present such information to the proper authority for action.

a. An Engineer in private practice will not review the work of another engineer for the same client, except with the knowledge of such engineer, or unless the connection of such engineer with the work has been terminated.

b. An Engineer in governmental, industrial or educational employ is entitled to review and evaluate the work of other engineers when so required by his employment duties.

c. An Engineer in sales or industrial employ is entitled to make engineering comparisons of his products with products by other suppliers.

Section 13—The Engineer will not associate with or allow the use of his name by an enterprise of questionable character, nor will he become professionally associated with engineers who do not conform to ethical practices, or with persons not legally qualified to render the professional services for which the association is intended.

a. He will conform with registration laws in his practice of engineering.

b. He will not use association with a nonengineer, a corporation, or partnership, as a "cloak" for unethical acts, but must accept personal responsibility for his professional acts.

Section 14—The Engineer will give credit for engineering work to those to whom credit is due, and will recognize the proprietary interests of others.

a. Whenever possible, he will name the person or persons who may be individually responsible for designs, inventions, writings, or other accomplishments.

b. When an Engineer uses designs supplied to him by a client, the designs remain the property of the client and should not be duplicated by the Engineer for others without express permission.

c. Before undertaking work for others in connection with which he may make improvements, plans, designs, inventions, or other records which may justify copyrights or patents, the Engineer should enter into a positive agreement regarding the ownership.

d. Designs, data, records, and notes made by an engineer and referring exclusively to his employer's work are his employer's property.

Section 15—The Engineer will cooperate in extending the effectiveness of the profession by interchanging information and experience with other engineers and students, and will endeavor to provide opportunity for the professional development and advancement of engineers under his supervision.

a. He will encourage his engineering employees' efforts to improve their education.

b. He will encourage engineering employees to attend and present papers at professional and technical society meetings.

c. He will urge his engineering employees to become registered at the earliest possible date.

d. He will assign a professional engineer duties of a nature to utilize his full training and experience, insofar as possible, and delegate lesser functions to subprofessionals or to technicians.

e. He will provide a prospective engineering employee with complete information on working conditions and his proposed status of employment, and after employment will keep him informed of any changes in them.

Note: In regard to the question of application of the Code to corporations vis-a-vis real persons, business form or type should not negate nor influence conformance of individuals to the Code. The Code deals with professional services, which services must be performed by real persons. Real persons in turn establish and implement policies within business structures. The Code is clearly written to apply to the Engineer and it is incumbent on a member of NSPE to endeavor to live up to its provisions. This applies to all pertinent sections of the Code.

NSPE Publication No. 1102 As Revised, January 1974

Appendix E

Court Deposition

GALL, LANE & POWELL
Frank Cummings
Jill Cummings
1250 Connecticut Avenue, N.W.
Washington, D. C. 20036
(202) 659-1600

ROBERT G. WERNER
1255 Post Street, (Suite 700)
San Francisco, California 94104
(415) 441-1211

Attorneys for The Institute of Electrical and Electronics Engineers, Inc.

FILED
JAN - 9 1975
RENE C. DAVIDSON, County Clerk
By [signature]
JAMES KITTERMAN, Deputy

SUPERIOR COURT OF CALIFORNIA

FOR THE COUNTY OF ALAMEDA

HOLGER HJORTSVANG, Plaintiff vs. SAN FRANCISCO BAY AREA RAPID TRANSIT DISTRICT, a public entity; DOES ONE through TEN INCLUSIVE, Defendants	NO. 436443 (Consolidated with Nos. 436444 and 436445)

AMICUS CURIAE BRIEF OF THE INSTITUTE OF ELECTRICAL AND ELECTRONICS ENGINEERS, INC.

I. STATEMENT

This brief is filed as amicus curiae because, on the basis of the pleadings, it is clear that rulings in this case will involve important questions concerning the proper ethics of an engineer in the employ of a public employer.

The Institute of Electrical and Electronics Engineers ("IEEE") is the largest engineering society in the nation and has a direct concern with the establishment, maintenance, and recognition (including governmental and judicial recognition) of ethics within the engineering field.

This brief is submitted with two limited aims: first, to inform this Court of the existence and terms of established standards and codes of ethics for engineers, in the employment context generally and particularly in the context of public employment;* and, second, to seek the Court's recognition that such standards and codes are relevant and material to this case for the reasons discussed below.**

II. SUMMARY OF ARGUMENT

This Court is expected to rule, as the trial proceeds, on questions of law, and this amicus curiae brief is addressed solely to those rulings.

Within that framework, we urge this Court to rule:

1. As to Admissibility of Evidence: That evidence of professional ethics of engineers, as outlined herein and as further developed by the parties, is relevant, material, and admissible;

2. As to Any Motions for Judgment: That, in consideration of any motion to dismiss or for judgment by this Court, the

---

*IEEE, moreover, is familiar with and can supply expert evidence concerning the ethical codes of engineers.

**IEEE takes no position on the merits and the claims, as IEEE has no direct evidence to offer as to what the claimants did, what defendants did, or why.

Court should rule that an engineer is obligated to protect the public safety, that every contract of employment of an engineer contains within it an implied term to the effect that such engineer will protect the public safety, and that a discharge of an engineer solely or in substantial part because he acted to protect the public safety is a breach of such implied term; and

3. As to Jury Instructions: In any charge to the jury herein, this Court should instruct the jury that if it finds, based upon the evidence, that an engineer has been discharged solely or in substantial part because of his bona fide efforts to conform to recognized ethics of his profession involving his duty to protect the public safety, then such discharge was in breach of an implied term of his contract of employment.

We base this position upon the cases, statutes and ethical codes discussed below.

POINT I

PROFESSIONAL ETHICS ARE MATERIAL AND RELEVANT

California judicially recognizes that an employee may not be arbitratily discharged where the discharge would be inconsistent with the public good, even if his employment contract is terminable at will. In Petermann v. International Brotherhood of Teamsters, 174 Cal. App. 2d (1959), it was held that an employer may not discharge an employee because the employee refuses to commit perjury. The public has too great a stake in the integrity of the judicial process to permit such a discharge.*

---

*See also Slochower v. Board of Higher Education of the City of New York, 350 U.S. 551 (1956).

3.

In *Petermann*, the District Court of Appeal for the Second District noted that the contract of employment did not provide for any fixed period of duration and that such a relationship is generally terminable at will, "for any reason whatsoever". But it also noted that such a right of discharge "may be limited by statute" or "by considerations of public policy". The Court then said at page 188:

> "By 'public policy' *is intended that principle of law which holds that no citizen can lawfully do that which has a tendency to be injurious to the public or against the public order* ...." (emphasis by the Court).

The Court then noted that, because the State had a declared policy against perjury, "the civil law, too, must deny the employer his generally unlimited right to discharge an employee whose employment is for an unspecified duration, when the reason for the dismissal is the employee's refusal to commit perjury." The Court said that "the law must encourage and not discourage truthful testimony. The public policy of this state requires that every impediment, however remote to the above objective, must be struck down when encountered." *Id.* at 188, 189. The lower court having dismissed, the Court of Appeal reversed.

When questions of public safety are at stake, an engineer's code of ethics stands in the same position as the laws against perjury. If a code of ethics properly requires the protection of the public, a discharge because an employee insisted on following that code would be inconsistent with the public good. Thus compliance with such a code must be deemed an *implied term* of the employment contract.*

---

*This court may, but need not, decide the extent to which the principles of this case would be applicable in the case of a private employer. The complaint in this case alleges that a public employer discharged public employees because those employees informed the public of a danger to the public safety. In a very real sense, the public at large was the "employer" of the plaintiffs herein; whatever may be the limits of the duties of public disclosure by the engineer in private employment, there is clearly a higher duty in the case of public employment.

California statutes clearly recognize an engineer's obligation to protect the public. California Government Code, Section 835 waives the State's sovereign immunity and makes a public entity liable for conditions dangerous to the public. Section 840.2(b) of the same Code makes a public employee liable if he fails to take adequate measures to protect the public from such conditions. That section obviously encompasses any and all engineers engaged in public employment.

The same recognition is reflected in California statutes governing licensing* of professional engineers, including electrical and mechanical engineers. California Business and Professional Code Section 6730 states that the purpose of that Code is "to safeguard life, health, property and public welfare." And Section 6775 provides that a licensed engineer may be disciplined -- indeed his registration may be revoked -- for "negligence", "incompetency in his practice", or if he "has not a good character".

What is "negligent", under ordinary common law principles, is determined by the scope of the negligent person's duties, and those duties are in part determined by what is generally recognized to be ethical. "Incompetency in his practice" involves failure to adhere to generally accepted standards of conduct and must be taken to include ethical standards, if those standards are widely publicized and generally recognized. And, most important, the notion of "good character", particularly in a professional sense, certainly involves adherence to generally accepted ethical standards, and particularly standards of professional ethics.

---

*Not all members of IEEE or other professional engineering societies are (nor are they all required to be) licensed to practice engineering in their home states. The ethical standards covering both licensed engineers and other engineers are the same, and this is particularly true where both types of engineers are working together on the same project, as was the case, we understand, in the BART situation.

California law, then, mandates adherence to ethical and moral standards. Engineers have adopted (see Point II below) proper ethical codes to complement statutory codes. We urge this Court on the Petermann principle to recognize (1) that an engineer has an overriding duty to protect the public, and (2) that California law, including statutes and case law, supports the drafting of ethical codes, makes the terms of generally accepted professional ethics relevant and material in a case such as this, and effects a legally enforceable incorporation of such codes into engineering contracts of public employment, insofar as such codes are widely acknowledged to be necessary for the protection of the public.

POINT II

ENGINEERING PROFESSIONAL CODES REQUIRE PROTECTION OF THE PUBLIC

1. A Common Thread: The Duty to Protect the Public.

The various professional engineering societies have, for many years, adopted and published Codes of professional ethics. Such codes contain at least one common thread -- that the engineer owes an overriding duty to protect the public safety.

For example, the Canons of Ethics for Engineers was prepared and adopted by the Engineers' Council for Professional Development ("ECPD") in 1946.* These Canons were then adopted by the Board of Directors of the National Society of Professional Engineers ("NSPE") in October 1946, and were published in NSPE's Journal, "The American Engineer", in its November 1947 issue.

---

*ECPD is an organization founded by a group of professional engineering societies, whose participants and affiliates now include the American Institute of Aeronautics and Astronautics, the American Institute of Chemical Engineers, the American Institute of Industrial Engineers, the American Institute of Mining, Metallurgical and Petroleum Engineers, the American Nuclear Society, the American Society of Agricultural Engineers, the American Society of Civil Engineers, the American Society for Engineering Education, the American Society of Mechanical Engineers, the Institute of Electrical and Electronics Engineers, National Council of Engineering Examiners, the Society of Automotive Engineers, National Institute of Ceramic Engineers, and the National Society of Professional Engineers.

Section 4 of these Canons provided:

> "He [the engineer] will have due regard for the safety of life and health of public employees who may be affected by the work for which he is responsible."

This code has an even longer history, having been discussed initially in the May 1935 issue of "The American Engineer", although the code was formally adopted in 1946, in a form differing little from the present code. *

NSPE's own code of ethics (distinct from ECPD's) was adopted in 1964, and published in the September 1964 issue of "The American Engineer".** This code provided, in Section 2:

> "Section 2 - The Engineer will have proper regard for the safety, health, and welfare of the public in the performance of his professional duties. If his engineering judgment is overruled by nontechnical authority, he will clearly point out the consequences. He will notify the proper authority of any observed conditions which endanger public safety and health.
>
> a. He will regard his duty to the public welfare as paramount.
>
> b. He shall seek opportunities to be of constructive service in civic affairs and work for the advancement of the safety, health and well-being of his community.
>
> c. He will not complete, sign, or seal plans and/or specifications that are not of a design safe to the public health and welfare and in conformity with accepted engineering standards. If the client or employer insists on such unprofessional conduct, he shall notify the proper authorities and withdraw from further service on the project."

We emphasize in this regard the code's injunction to the engineer that he must "notify the proper authority" of anything he observes which may "endanger public safety". We think it fair to say that the ultimate proper authority in the case of public employment is the public itself.

---

*The ethical proposal originally published by NSPE in the May 1935 issue of "The American Engineer" included the following: "The engineer shall at all times and under all conditions seek to promote the public welfare by safeguarding life, health and property."

**NSPE, when it published its code in 1964, had membership of 62,038 engineers, and its journal was circulated, in addition, to over 1,000 libraries and institutions. Its membership today is approximately 70,000 engineers.

ECPD, meanwhile, adopted revised Canons in September 1963, which stated, in the very opening paragraph:

> "1.1 -- The Engineer will have proper regard for the safety, health and welfare of the public in the performance of his professional duties."

These Canons were adopted by a variety of professional engineering societies. The American Society of Mechanical Engineers, whose membership now totals close to 70,000, ratified these canons in 1963, and they were published in ASME's magazine, "Mechanical Engineering".

The same principles are carried forward to the current day. For example, a set of "Guidelines to Professional Employment of Engineers and Scientists" published by the IEEE Board of Directors in its national monthly magazine, *Spectrum*, in April, 1973,* contains the following paragraph:

> "The professional employee should have due regard for the safety, life, and health of the public and fellow employees in all work for which he/she is responsible. Where the technical adequacy of a process or product is involved, he/she should protect the public and his/her employer by withholding approval of plans that do not meet accepted professional standards and by presenting clearly the consequences to be expected if his/her professional judgment is not followed."

*A much earlier code, adopted and published by the American Institute of Electrical Engineers (IEEE's predecessor) in 1912 provided: "An engineer should consider it his duty to make every effort to remedy dangerous defects in apparatus or structures or dangerous conditions of operation, and should bring these to the attention of his client or employer." The "employer", in a case such as this, is first the public entity and ultimately the California general public which is the entity's own employer. IEEE supplemented the 1912 code in 1974 by a new code which includes the following: "Engineers shall, in fulfilling their responsibilities to the community: (1) protect the safety, health and welfare of the public and speak out against abuses in these areas affecting the public interest...."

2. General Acceptance and Publication of the Common Thread.

Because the cited codes have been widely circulated and generally endorsed, it seems eminently reasonable to conclude that every engineer is aware of his obligation to the public. The guidelines published by IEEE, for example, have also been endorsed by over twenty societies.*

Even before the engineer's obligation to serve the public was fully codified in writing, moreover, there was an historical recognition of that obligation, discussed in professional journals.**

CONCLUSION

Based upon the foregoing, we submit and we urge this Court to acknowledge that an engineer has an overriding obligation to protect the public.

Specifically, we urge this Court:

(1) To rule that evidence of professional ethics is relevant, material and admissible in this case; and

(2) To rule, as to any motions for judgment or any jury instructions, that an engineer is obligated to protect the

---

*The endorsing societies include: American Association of Cost Engineers, American Institute of Aeronatics and Astronautics, American Institute of Chemical Engineers, American Institute of Chemists, American Institute of Industrial Engineers, American Institute of Professional Geologists, American Nuclear Society, American Society of Agricultural Engineers, American Society of Engineering Education, American Society of Civil Engineers, American Society of Mechanical Engineers, American Society of Quality Control, Data Processing Management Association, Engineering Societies of New England, Inc., Engineers Council for Professional Development, Engineers Joint Council, Institute of Electrical and Electronics Engineers, Instrument Society of America, Institute of Traffic Engineers, National Association of Corrosion Engineers, National Institute of Ceramic Engineers, National Society of Professional Engineers, Society for Technical Communications, Society for Experimental Stress Analysis, Society of Fire Protection Engineers, Society of Women Engineers, Technical Association of the Pulp & Paper Industry.

**The code of ethics of the NSPE, for example, was discussed initially in the May, 1935, issue of the American Engineer although that code was first formally adopted in 1946 (in a form differing little from the present code.).

public safety, that an engineer's contract of employment includes as a matter of law, an implied term that such engineer will protect the public safety, and that a discharge of an engineer solely or in substantial part because he acted to protect the public safety constitutes a breach of such implied term.

DATED: January 9, 1975.

Respectfully submitted,

GALL, LANE & POWELL
1250 Connecticut Avenue N.W.
Washington, D.C. 20036
(202) 659-1600

By Frank Cummings
Frank Cummings

By Jill Cummings
Jill Cummings

ROBERT G. WERNER
1255 Post Street (Suite 700)
San Francisco, California 94109
(415) 441-1211

Robert G. Werner

Attorneys for The Institute of Electrical and Electronics Engineers

Appendix F

BART Chronology

January, 1947 A Joint Army-Navy Review Board recommends the construction of an underwater transit tube beneath San Francisco Bay linking San Francisco with Oakland, CA. Hope was expressed that this connecting link would prevent intolerable congestion on the Bay Bridge.

July 25, 1951 The California State Legislature creates the 26-man San Francisco Bay Area Rapid Transit Commission with representatives from each of the nine counties which touch the Bay. The Commission is to study long-range transit needs and recommend the best solution to future problems.

January, 1953 The San Francisco Bay Area Rapid Transit Commission makes a preliminary report to legislature about a proposed rapid transit plan.

November, 1953 The San Francisco Bay Area Rapid Transit Commission retains Parsons, Brinckerhoff, Hall & MacDonald (PBHM) of New York as the prime contractor to conduct a broad range of studies.

January, 1956 Parson, Brinckerhoff, Hall & MacDonald recommends the construction of a regional rapid transit system as a solution to the Bay area traffic congestion problem.

February, 1956 Stanford Research Institute recommends the creation of a public district to plan, construct, and operate a Bay Area Transit system.

January 17, 1957 The San Francisco Bay Area Rapid Transit Commission makes its final report to the California State Legislature. It recommends forming a five-county rapid transit district "to build and operate a high- speed rail network linking major commercial centers with suburban centers."

June 4, 1957 Acting on the San Francisco Bay Area Rapid Transit Commission's recommendations, the California State Legislature forms the San Francisco Bay Area Rapid Transit District (comprising Alameda, Contra Costa, Marin, San Francisco, and San Mateo counties) to plan and, if approved, to build and operate a regional rapid transit system.

November 14, 1957 The Bay Area Rapid Transit District (BARTD) holds its first meeting with representatives of Alameda, Contra Costa, Marin, San Francisco, and San Mateo counties. BARTD is initially governed by a 16-member board of directors appointed by Boards of Supervisors and mayoral committees within the 5 counties.

January 1, 1958 The first BARTD office is opened in San Francisco.

May 14, 1959 BARTD retains 3 engineering firms (Parsons, Brinckerhoff, Hall & MacDonald; Tudor Engineering Co.; and Bechtel Corp.) to develop a regional transit plan.

July 10, 1959 The California State Legislature passes a bill approving the use of San Francisco Bay Bridge tolls for the construction of an underwater rapid transit tube. This was one of the original recommendations of the Army-Navy Review Board in 1947.

September 10, 1959 BARTD retains Ebasco Services, Inc. to perform economic studies pertaining to rapid transit in the Bay Area.

October 8, 1959 BARTD retains Smith, Barney & Co. of New York to develop a financial plan for a Bay Area Rapid Transit (BART) system.

January 20, 1960 The California State Legislature approves the use of the Grove-Shafter Freeway median for a BART route.

March-April, 1960 BARTD holds public hearings on its developing rapid transit plan in the 5 counties where it is to be constructed.

July 1, 1960 Engineering consultants complete a feasibility study of the "Transbay Tube," the underwater rapid transit tube to be placed beneath the San Francisco Bay.

September 1, 1960 BARTD sends a <u>tentative</u> rapid transit plan to city and county officials for comments.

February 9, 1961 BARTD directors approve a five-county rapid transit route.

June, 1961 Parsons, Brinckerhoff, Tudor, & Bechtel (PBTB) submits a <u>final</u> plan for a five-county system, which it estimates will require 10 years to build at a cost of $1.2 billion. Financial consultants recommend that BARTD seek voter approval for $1.08 billion in general obligation bonds (plus $133 million from Bay Bridge tolls) to pay for the construction.

June 6, 1961 The California State Legislature passes a bill setting a 60% vote requirement for authorization by the electorate of the BARTD transit plan. This is a change from the customary two-thirds (66%) majority required to pass a bond issue.

August 1, 1961 Golden Gate Bridge directors reject rapid transit operation the on Golden Gate Bridge.

September, 1961 A comprehensive engineering, financial, and economic report prepared by Stone & Youngberg makes a good statistical case for the construction of the 5-county system to relieve the predicted commuter congestion in the Bay Area.

December 19, 1961 Citing the high cost of the system and the existing Southern Pacific Commuter trains, San Mateo County Supervisors <u>vote</u> to withdraw their county from BARTD.

April 12, 1962 San Mateo County <u>officially</u> withdraws from BARTD.

April 17, 1962 A PBTB Composite Report predicts the cost of the basic BART system, not including the Transbay Tube & rolling stock, to be $792 million. The Transbay Tube is estimated to cost $133 million plus $73 million for 450 rapid transit cars. The total cost estimate is $988 million.

May 17, 1962 Marin County withdraws from BARTD citing as reasons the inability of the Golden Gate Bridge to carry transit vehicles and the prohibitive cost of the BART system.

May 24, 1962 BARTD directors adopt a three-county rapid transit plan and formally transmit it to the Alameda, Contra Costa, & San Francisco County Boards of Supervisors.

July 9-24, 1962 Boards of Supervisors of remaining the three BARTD counties (San Francisco, Alameda, and Contra Costa) approve the three-county BARTD plan. The plan is then placed on the ballot for the following general election in November.

November 6, 1962 Voters of the three counties approve a $792 million Bay Area Rapid Transit plan and authorize construction of the 71.5 mile system, consisting of 33 stations in 17 communities in the 3 counties.

The plan wins in San Francisco with 68% of the vote, 60% in Alameda, & 54% in Contra Costa. The average total vote for approval was 61.2%.

The total cost of the system is projected, in 1962, as $996 million ($792 million for 71.5 miles; $133 for the Transbay Tube to come from bonds issued by the California Toll Bridge Authority; $71 million for rolling stock to be funded primarily by bonds issued against future operating revenues).

(NOTE: Estimates of the cost of the BART system differ somewhat. The difference between a $988 million estimate and a $996 million estimate, however, is only a fraction of 1%.)

November 29, 1962 BARTD signs a contract with Bechtel Corp., Tudor Engineering Co., and Parsons, Brinckerhoff, Quade and Douglas (PBTB) to engineer and manage construction of the rapid transit system.

However, a taxpayers' suit 7 days after the bond issue delays the start of work for 6 to 9 months. The suit, filed in Contra Costa County, challenged the California State Legislature's authority to reduce the minimum required vote percentage, the involvement of BARTD officials in the referendum campaign, the contract with BARTD's engineering consultants, and the legality of BARTD itself.

December 8, 1962 The Ethical Practices Committee of the California Society of Professional Engineers (CSPE) holds an all-day review of the BARTD/PBTB contract and recommends that BARTD form a "Board of Advisors, experienced in the employment of Engineering services," to reevaluate the contract.

1963 B. R. Stokes, a BARTD public relations man, is promoted to General Manager of BARTD.

June 10, 1963 Contra Costa Superior Court rules in BARTD's favor on all points raised in the taxpayers' suit challenging the validity of the bond election.

BARTD receives an initial $4.88 million Federal mass transit demonstration grant to be applied to a test program.

July, 1963 Full-scale design engineering for BART is begun by PBTB.

November 27, 1963 Right-of-way acquisition for BART begins.

1964 David Hammond joins BARTD as Assistant General Manager, Operations & Engineering. Hammond is a retired Colonel in the Army Corps of Engineers who has been in charge of all Army construction in the continental U.S.

C. O. Kramer joins BART as superintendent of Power & Way. He previously served 3 years as Chief Engineer of the Southern Pacific Co.

June 19, 1964 BART construction officially begins. President Lyndon Johnson officiates at the Diablo Test Track groundbreaking at Concord, CA.

Late 1964 Work begins on one of the most difficult engineering projects of the entire BART system, construction of 5-Km twin-tunnels through the Berkeley Hills.

1965 Budd Co. of Philadelphia builds 3 stainless-steel "shell" cars that are used to test comfort, seating arrangement, & styling for maximum comfort & passenger-carrying capacity on the Diablo Test Track.

April 12, 1965 The Diablo Test Track is placed in operation between Walnut Creek & Concord, CA in Contra Costa county. The test track is used until February, 1966.

Cost of the test is $7.3 million; $4.9 million comes from a U.S. Housing & Home Finance Act of 1958 grant given to BARTD on June 10, 1963.

Summer 1965 BARTD admits that costs are beginning to exceed design estimates by a wide margin. The BARTD staff assumes cost control of the project. Some of the PBTB consultants fear that the over emphasis on the "Economy Kick" will militate against esthetics & sound engineering. The California State Legislature, disturbed by reports of a financial crisis, welcomes the shift of financial responsibility from the PBTB consultants to the BARTD staff.

December, 1965 The BARTD staff recalls this as "Black Christmas". Only 2 bids are received on the Oakland Subway. The lowest bid exceeds the engineers' cost estimates by $35 million.

(week later): The low bid on the Transbay Tube overruns the designer's estimates by almost $30 million (23%).

February 10, 1966 BARTD rejects all bids on the downtown Oakland Subway and directs the repackaging of the contract. PBTB is given 6 months to redesign the Subway.

Early 1966 It is evident by this time that construction costs could increase by as much as $182 million (18%) & result in a financial deficit of about $150 million (15%).

April, 1966 BARTD awards a $90 million contract for construction of the Transbay Tube.

August, 1966 PBTB issues specifications for the Automatic Train Control (ATC) system.

August 25, 1966 BARTD receives approval from H.U.D. for a $13.1 million capital construction grant. The total sum received from federal grants exceeds $159 million or 16% of the orginal BART cost estimate.

September 8, 1966 BARTD Board of Directors receives a project reestimate and adopts a policy statement on long-range financing.

September 12, 1966 Holger Hjortsvang is hired by BART as a Train Control Engineer. He is the "first Train Control Engineer" hired.

October 4, 1966 Berkeley voters approve a Special Service District to finance a border-to-border subway in their city.

November, 1966 Construction begins on the 3.8 mile Transbay Tube.

1967 The California Society of Professional Engineers (CSPE) claims a BART employee was transferred from auditing after making a number audits critical of BARTD's handling of funds.

February-March 1967 BARTD awards a $26 million contract to Westinghouse Electric Co. to design, install, and operationally qualify the ATC system. Various sources list the exact date of this contract as February 28, March 23, and March 31. (By 1974, the total estimated cost of the ATC system was $40 million, a 65% increase.)

July 25, 1967 Construction begins on the Market Street subway and stations beneath the city of Francisco. [Market Street was disrupted sectionally for 6 years as excavation progressed southward from the Bay end.]

September 14, 1967 BARTD reaffirms its policy to obtain full financing for the entire system before beginning revenue operation.

1968 Westinghouse strongly recommends BART engineers be assigned to their Pittsburgh office to follow the development of the ATC system & to work with Westinghouse engineers.

Ten-month-period in 1968-1969 Holger Hjortsvang is assigned by BARTD to the Westinghouse facility in Pittsburgh, PA.

Late 1968 Insufficient funds remain from the original bond issue of 1962 for the completion of right-of-way & purchase of BART vehicles.

November, 1968 A Department of Transportation (DOT) grant for $28 million is received for the development & purchase of rolling stock.

Following this action, Governor Ronald Reagan calls the California State Legislature into 2 special sessions to find a means to raise additional money for BARTD.

1969 The Diablo chapter of CSPE argues that increased delays & cost overruns will occur if BARTD does not change its policies. They call for a full state investigation of BARTD, but no action is taken.

Engineering personnel begin calling BART management's attention to deficiencies in BART.

March 28, 1969 The California State Legislature approves a 1/2 cent District sales tax to provide $150 million to complete the system.

April 2, 1969 The first of Holger Hjortsvang's memos questioning the BARTD operation is sent.

July, 1969 First projected opening date of the full BART system (according to California State Senator John Nejedly).

July 3, 1969 BARTD awards a $67 million transit vehicle contract for 250 cars to Rohr Corp., an aerospace industry manufacturer in Chula Vista, California.

August, 1969 The Transbay Tube structure is complete.

When the ATC contract was awarded in early 1967, the initial segment from MacArthur to Hayward was scheduled to open in August, 1969.

September 19, 1969 Roy Anderson becomes the sixth announced candidate for appointment to the BARTD Board of Directors representing Contra Costa county. Anderson is sponsored by the Diablo chapter of CSPE.

September 23, 1969 Gil Verdugo places Roy Anderson's name in nomination for the BARTD Board at the Contra Costa Board of Supervisors meeting.

September 24, 1969 The Contra Costa Board of Supervisors reappoints Joseph Silva and elects Nello Bianco to seats on the BARTD Board of Directors.

November 10, 1969 Robert Bruder is hired by BARTD as an Electrical/Electronic Engineer.

February, 1970 BARTD joins with Oakland, Alameda County, & the Coliseum to study the feasibility of linking the Coliseum Station to Oakland Airport.

April, 1970 BARTD officials work with San Francisco & San Mateo Counties to develop plans for extending BART from Daly City southward to the San Franciso International Airport.

June, 1970 The Southern Alameda County Line is energized & laboratory car testing begins.

August 28, 1970 The first 10 prototype cars arrive & test operations begin on the Southern Alameda County Line.

After delivering the first 10 prototype cars, Rohr starts construction of 250 production (revenue service) cars.

October, 1970 IBM demonstrates a prototype of the Automated Fare-Collection System.

Late 1970 Full operation of the BART system is scheduled in the original contracts for this time.

Early 1971 Opening of the Transbay Tube is initially scheduled for this time.

February 5, 1971 The last of Hjortsvang's memos questioning BART operation is sent.

March, 1971 Dr. Woodrow Johnson is given the BART assignment by Westinghouse "when the system's test program was in deep trouble."

March 25, 1971 Another $40 million grant is received from DOT for rolling stock.

May, 1971 According to the Burfine Report (see December 6, 1971), until May, 1971, little attention was devoted to checking out the computer portion of the system by PBTB.

May 10, 1971 Max Blankenzee is hired by BART as a Senior Programmer/Analyst.

June 11, 1971 Original "circulation" deadline for prototype cars (delayed until November, 1971).

June 14, 1971 Target date for completion of the ATC system is missed by Westinghouse.

July 1, 1971 Initial desired opening day for the BART system with 150 cars operating.

July 23, 1971 The last rail is set into place on the Contra Costa Line to complete the linking of all the system's mainline trackage.

August, 1971 Westinghouse attempts to conduct static testing of ATC software with a simulator and fails repeatedly.

September 20, 1971 The Battelle Institute report is issued. It points out a train-detection problem.

September 8,9,10, 1971 Blankenzee & Hjortsvang attend a seminar given by the Advance Institute of Technology in Standards in Program Development.

Sept.22,1971 - Nov.5,1971 Blankenzee sends his supervisor, Ed Wargin, a series of memos criticizing aspects of the ATC system.

October 18, 1971 The first successful test of ATC software is conducted.

Blankenzee tries to locate "BART's operating philosophies," but finds they aren't developed yet.

October, 1971 Daniel C. Helix becomes a member of the BART Board of Directors.

November, 1971 Delayed deadline for "circulation" of prototype cars.

San Francisco Airport access study is completed and possible extension studies for service east of Concord on the Contra Costa Line begins.

Blankenzee speaks with BARTD Director William Blake by phone. Blankenzee expresses his desire to speak with upper management. Blankenzee maintains that Blake told him to set up a meeting which he would attend. Blankenzee is unable to reach Blake again, and a meeting is never held.

November 2, 1971 Two cars are damaged in a collision at the Coliseum Station.

November 5, 1971 The first production car for revenue service is delivered by Rohr Corp.

November 10, 1971 A Board of Inquiry report on the November 2 collision is forwarded to all BARTD directors. The Maintenance Dept. recommends that the cars be scrapped.

November 18, 1971 An anonymous memo (from Hjortsvang) critical of BARTD Management circulates among BART's Technical Staff.

November 29, 1971 A strike at the Rohr plant (which continued until February 1, 1972) forces Rohr to ask for a 210 day extension of its delivery dates (BART, PBTB & Rohr settle on 150 days). The Rohr plant is shut down for 9 weeks and it is estimated that a total of 36 weeks of production time is lost.

December, 1971 The new BARTD Headquarters at Lake Merritt Station in Oakland are dedicated. All BARTD activities are transferred to this location from the temporary offices in San Francisco.

A moving BART train hits a stationary train. A Board of Inquiry is convened by Stokes. The BARTD Board of Directors is not satisfied with the results of the inquiry and tells Stokes to reconvene the Board of Inquiry. It is never done.

Edward Burfine, a consultant from Beckers, Burfine & Associates, spends part of a day examining the BART system and listening to the views of the three engineers (Hjortsvang, Bruder & Blankenzee). He writes what comes to be known as "The Burfine Report."

According to Hjortsvang-- Blankenzee, Hjortsvang, Helix & Gilbert Ortiz (a union organizer) meet at Union Headquarters in Oakland, CA. The meeting is reportedly arranged by Ortiz. Hjortsvang gives Helix a copy of his November 18, 1971 memo.

At a Board of Directors' meeting, Helix asks Hammond if there are any problems with the ATC system or with Westinghouse. Hammond says there are nothing other than routine-type problems; he implies the only major problem is lack of cars caused by a strike at the Rohr plant.

December 29, 1971 Bill Bugge and Jack Chambers of PBTB tell Helix there are significant problems with the ATC system because of Westinghouse's failure "to put their top people on the BART job." Prior to this meeting, Helix gives Justin Roberts, a reporter from the Contra Costa Times, a copy of the Burfine Report but asks him not to print it until he speaks to Stokes. Helix considers this meeting a "snow job" and allows Roberts to release the report.

Late 1971 The East Bay Line from Fremont to MacArthur Station, Oakland is originally scheduled to begin service. It is delayed by the inability of Rohr to deliver cars.

January, 1972 Articles appear in the <u>Contra Costa Times</u> concerning problems with the BART ATC system.

Blankenzee tries to stop PBTB from running total acceptance tests of the electrification system because the subsystems aren't approved. He claims that he is told he is "all wet."

General Manager Stokes makes public statements that BART will be open in April, 1972. (Testing on "crucial aspects of the train control system" does not even begin until May, 1972.)

At the January BARTD Board of Directors Administrative Committee Meeting, Hammond acknowledges problems with Westinghouse relative to the Train Control Contract. Helix claims Hammond doubts there is anything the Board of Directors can do about the problems. Helix tells him that it is a matter for the Board to decide.

Early 1972 BART is scheduled to open. It does not.

January 10, 1972 Helix sends a confidential memorandum to the President and Vice-President of the BARTD Board of Directors indicating that he has met with some employees who are critical of BART.

January 12, 1972 A seven-page report entitled "Review of BART operations" ("The Burfine Report") prepared by Edward Burfine is delivered to the BARTD Board of Directors. The report is based upon Burfine's December, 1971 visit to BART which he claims was requested by two members of the BARTD Board of Directors. The two directors (Nello Bianco and William Blake) deny hiring Burfine.

January 27, 1972 BARTD Board of Directors' President George Silliman orders the Board of Directors' Engineering Committee (composed of Directors Wm. Reedy, chairman, Arnold Anderson, Daniel Helix, and Joseph Silva) to investigate the ATC system because of claims made in the Burfine Report and because of articles in the <u>Contra Costa Times</u> with Helix as a source of negative information about BARTD.

January 31, 1972 BARTD Director Wm. M. Reedy, Engineering Committee chairman, directs management to obtain information on Beckers, Burfine & Associates, and find out who hired them.

February, 1972 The first revenue vehicles are received after the Rohr strike ends.

Blankenzee is called into Wargin's office and denies knowing who contacted Burfine.

The next day, Blankenzee is called into E. J. Ray's (BARTD Director of Operations) office and denies knowing about Burfine.

Bruder is called into John B. Fendel's (BARTD Chief of Construction) office and denies knowing Burfine.

Hjortsvang refuses to answer Ray's questions about who hired Burfine.

February 22, 1972 The BARTD Board of Directors Engineering Committee Meeting is attended by 11 of 12 Board members. Dr. Woodrow Johnson (Vice President of Westinghouse) presents a report claiming "All our major problems are behind us." John R. Asmus, PBTB manager of engineering, reports that BART is 30% complete. Edward Burfine, of Beckers, Burfine and Associates, presents a report which is critical of BART. Helix's motion that Burfine or a similar firm be employed to pursue, with the General Manager and staff, questions on the ATC system for a report to the Board is defeated.

Helix says he will ask some concerned BART engineers to speak to the Board.

Helix, Bruder, Hjortsvang, and Blankenzee participate in a conference call in which they discuss if any of the engineers will appear before the BARTD Board of Directors to explain their views on the ATC system which had been discussed at the Engineering Committee meeting earlier in the day.

February 23, 1972 Bruder calls Wm. Jones (CSPE President) early in the morning to express concern about his job. Within the hour, Jones contacts the Diablo chapter to which Bruder belongs. In the evening, the Diablo chapter Board meets with Bruder. [7:45 AM - Bruder calls Jones at home; 8:10 AM - Bruder leaves a message for Jones at Jones' office; Jones calls Gil Verdugo; Verdugo calls Bruder; 10:55 AM - Jones calls Jim Wright (of the Western Council of Engineers); 12:10 PM - Verdugo calls Jones; 1:35 PM - Jones tries to call Bruder; 7:30 PM - Meeting at Verdugo's office: Verdugo, Anderson, Bruder & Jim Riffell (secretary of the Diablo Chapter) attend.]

February 24, 1972 The BARTD Board of Directors votes 10 to 2 to accept the staff report submitted to the Engineering Committee Tuesday, February 22 confirming that the basic responsibility for the ATC system rests with Westinghouse and that problems, if any, will be corrected by Westinghouse. The Board closes the issue with this vote of confidence for Stokes. Helix and Blake vote against this endorsement of BARTD management. Gil Verdugo attends the meeting as an observer.

Bruder meets with some BART operations' employees. He claims that he casually asks them if they are willing to go before the Board of Directors; E. A. Tillman (BARTD Chief of Engineering and Construction) claims Bruder was "recruiting them."

February 28, 1972 Verdugo sends a letter to Jones with newspaper articles "covering the recent actions of the BARTD board." Verdugo notes, "We will continue to follow the situation with Bruder and hope to come up with some conclusion rather than close the subject as the BARTD board did. There may be an 'iceberg' here so we should gather facts before we act."

Late February Bruder visits Frank Wagner (one of his supervisors) and tells him that he talked with Helix, Hjortsvang & Blankenzee about going to the Board of Directors to explain their views on the ATC system. Wagner calls his supervisor, Fendel. Fendel tells Wagner to talk to Tillman.
Next day - Wagner tells Tillman.

March 1, 1972 Jones responds to Verdugo's letter of February 28 by noting, "I agree completely with the conclusion you draw matter are: 1. that having heard that some engineers think they have a problem, we (CSPE) are interested in resolving it; 2. that we take actions to resolve it, if indeed it exists; and 3. that the engineers know of CSPE's interests and actions."

Stokes issues a memo giving reasons why the 3 engineers should be fired.

3:00 PM - Stokes confers with E. J. Ray, C. O. Kramer (General Superintendent, Power & Way), and L. Kimball. He decides to discharge Hjortsvang & Blankenzee. Ray & Kramer are told to advise Hjortsvang & Blankenzee of the grounds for discharge.

March 2, 1972 9:30 AM - Hjortsvang and Blankenzee are called individually into Ray's office with Kramer present. Each is given the option to resign or be fired. Both refuse to resign & are fired. They are taken to their offices by a Security Division officer, collect their belongings, and are escorted out of the building.

Ray claims he notified the engineers of the grounds for their discharge.

March 3, 1972 E. R. Tillman sends a memo detailing Bruder's involvement in the "Burfine/Helix Matter."

11:40 AM - Tillman meets with Bruder & Fendel at the Lake Merritt Headquarters. Bruder denies involvement with Burfine and Helix. Tillman tells him to resign or be fired for lying. Bruder refuses to resign & is fired.

March 4, 1972 Gil Verdugo reports the firing of the three engineers to William F. Jones (CSPE President).

March 7, 1972 Stokes says the firing of the engineers is a "personnel matter."

March 9, 1972 Blankenzee calls Mark Bowers (BARTD Director of Employee Relations) requesting a copy of his discharge action and asking about the BARTD appeal procedure. He is told to direct his appeal to Ray. Blankenzee claims Ray never explained the reasons for his discharge.

March 10, 1972 Blankenzee writes to Ray asking for the procedures he should use to appeal his case.

March 12, 1972 The 3 engineers meet with CSPE representatives to decide what action CSPE should take. The meeting is at the Western Council of Engineers (WCE) headquarters in Oakland, CA. Bill Jones (CSPE President), Gil Verdugo (CSPE State Director and Diablo Chapter member), Roy Anderson (CSPE Diablo Chapter Director), Ron Tusgita (CSPE East Bay Chapter President), Jim Lackey (CSPE Diablo Chapter President), Jim Wright (Western Council of Engineers), Jim Walsh (NSPE National Director), and Holger Hjortsvang, Max Blankenzee, and Robert Bruder (the 3 fired engineers) are present.

Jones has dinner at Ed Walker's (President, CSPE Golden Gate Chapter) house and tells him about the meeting.

March 13, 1972 Jones tries unsuccessfully to contact Stokes. He speaks to Hammond who expresses surprise that CSPE is interested in the situation.

Later in the week, Jones calls Lou Riggs (president of Tudor) to try to set up a meeting with Stokes. Riggs calls Jones back to say that no meeting is possible. BARTD's top management declines to meet with CSPE.

March 13 - Daily, for 3 weeks, CSPE representatives interview people on the BART situation. The information they receive is described as "distressing."

March 14, 1972 Gil Verdugo, Jim Lackey, Ron Tsugita, Roy Anderson, George Humphrey (CSPE Diablo Chapter member), Justin Roberts (a reporter for the Contra Costa Times), Irwin Black (Roy Anderson's supervisor), Jim Walsh, Byron Tomlinson (CSPE Vice President), and Ed Walker meet at Verdugo's office in Lafayette, CA.

March 16, 1972
Hjortsvang, Verdugo, and Anderson meet.

March 17,1972 Jones, Walsh, Walker, Verdugo, Lackey, Anderson, Roberts & the 3 engineers meet.

March 18, 1972 Blankenzee, Verdugo, and Anderson meet.

March 19, 1972 Verdugo and Anderson start drafting a statement of the information they have accumulated from their previous meetings on the BARTD engineers' firing.

March 20, 1972 Anderson & Wright meet.

March 21, 1972 Anderson and Verdugo work on a draft of the report started on March 19. Bill Jones calls Anderson regarding his (Jones) discussion with Lou Riggs. (See March 13, 1972)

March 22, 1972
Verdugo & Anderson finish their report.

March 23, 1972 The Diablo chapter of CSPE meets and Jim Wright (of the WCE) attends. The Anderson and Verdugo report is presented & the chapter expresses support. The membership votes with one abstention to create a legal fund.

Jim Walsh volunteers to attempt to set up a meeting between representatives of CSPE, BARTD and PBTB. His offer is accepted by CSPE Diablo Chapter members, but he is unable to arrange the meeting. BARTD management refuses to attend on the grounds that CSPE has no appropirate role to play in BARTD's internal personnel decisions.

March 24,1972 Blankenzee receives a letter from Stokes saying an "independent investigation" of his discharge was conducted by BART's Department of Labor Relations & reviewed by top management; it was concluded that his discharge was warranted and his appeal is denied.

March 25, 1972
Verdugo, Bruder, Blankenzee, Hjortsvang, Tsugita, & Anderson meet. Blankenzee, Bruder and Hjorstvang are asked to prepare a statement on their dismissal action.

March 26, 1972
Anderson & Verdugo draft a statement of confidence in the 3 engineers. (43.113)

March 27, 1972 President of the CSPE Diablo chapter Jim Lackey writes to other CSPE chapters. He requests $5 per member for a legal case against BARTD.

April, 1972 Westinghouse is still having trouble with the ATC testing.

Early April, 1972 Jones receives the report written by Anderson and Verdugo (see March 26, 1972) and responds by notifying the CSPE attorney to take legal action against BARTD for firing the 3 engineers and to examine the safety issues involved.

Middle April, 1972 The 3 BARTD engineers, Gil Verdugo, Roy Anderson, Bill Jones, and one other member from the CSPE Diablo Chapter meet with State Senator John Nejedly at his Sacramento office. They explain the BARTD situation and Nejedly asks them to "get the facts together for a presentation to a group of legislators."

Jones cuations Verdugo and Anderson that "any implications adverse to the consulting engineering firm would be looked upon unfavorably and could not be supported by him."

Later, Verdugo, Anderson & Walker meet for dinner. According to Anderson, Walker is nervous and wants to schedule a meeting with the consulting engineers.

April 27, 1972 BARTD Board of Directors votes its first priority for revenues from State gasoline taxes to provide express bus feeder service to BART stations from areas in Contra Costa and Alameda Counties not served by public transit.

May 10, 1972 Verdugo, Anderson, Blankenzee and one of the other BARTD engineers go with Elliott Pearl (CSPE's attorney) to the offices of the BARTD attorneys. Only Pearl speaks to BARTD's attorneys. He reports that BARTD's attorney proposes a tentative settlement including letters of recommendation for the engineers and a CSPE-appointed technical advisory committee for BARTD. The attorney stresses that the offer is tentative and has not been cleared with BARTD management. Verdugo calls Jones to describe the preliminary tentative settlement offer. The offer is never pursued and nothing comes of it.

May, 1972 The state meeting of CSPE is held. The President's Award goes to the Diablo chapter in recognition of Anderson and Verdugo's work in supporting the 3 BARTD engineers. Jones urges full CSPE membership support for the 3 engineers.

Mid 1972 The BARTD Board of Directors sets September 11 as the first day of revenue service.

June, 1972 Bob Kuntz is installed as CSPE President replacing William Jones.

Senator Nejedly asks the Legislative Analyst's Office to investigate the safety and operation of BART.

250 cars were supposed to have been delivered by Rohr by this date, so far BART only has 81.

June 3, 1972 A BART train derails when a switch is activated under a moving train causing the rear wheels to derail. The train comes to rest against the third rail carrying 1000 volts.

June 5, 1972 CSPE writes to BARTD Board of Directors President Silliman and suggests starting service under manual control because of unresolved problems and recommends hiring an outside consultant to assure the system's safety & reliability.

WeeK of June The BARTD staff accepts the ATC system from Westinghouse.

June 8, 1972 BART pre-revenue train testing begins on the Southern Alameda Line.

Members of a CSPE Committee meet with the California Public Utilities Commission (PUC). PUC claims they wrote BARTD in May asking when the BART system would be ready for checking by PUC. They have not yet received notice from BARTD. PUC testing will take about 13 weeks & cannot start until the necessary information is filed.

June 9, 1972 A BART train traveling at 25 MPH under manual control fails to stop & runs through a closed switch even though the emergency brake is applied. Wet rails due to a light rainfall are cited as the cause of the accident.

June 13, 1972 Silliman replies to the CSPE letter of June 5th. He claims "All is well" & appreciates CSPE's interest. He reports that the BART system is safe.

June 18, 1972 Anderson, Roberts, and Jones meet at Blankenzee's home and review the events to date. Bill Jones asks Anderson not to send his report ("The BARTD Inquiry") to Senator Nejedly. Anderson phones Verdugo after the meeting. Verdugo creates a Transportation Safety Committee for the CSPE Diablo Chapter and appoints Anderson its chairman.

June 19, 1972 "The BART Inquiry" by Roy Anderson is submitted to Senator Nejedly with a cover letter signed by Anderson as Chairman of the Committee on Transportation Safety of the Diablo Chapter of CSPE

According to Roy Anderson, BARTD instructs its employees not to talk to CSPE.

Summer 1972 This is the initial target date for Transbay Tube operation.

The Golden Gate chapter of CSPE disagrees with the Diablo chapter's decision to back the 3 BARTD engineers and files a motion with the National Society of Profesional Engineers (NSPE) in Washington to censure any moves taken by the Diablo chapter. (Note: Hammond and some engineers from PBTB were members of the Golden Gate chapter.)

July, 1972 Justin Roberts, Gil Verdugo, and Roy Anderson meet with Senator Nejedly who claims that he is unable to get support for an investigation into BARTD.

A petition by the CSPE Diablo Chapter calling for the California State Legislature to conduct a BARTD investigation is drafted by Verdugo and Anderson. An emergency meeting of the Diablo chapter board is held; the petition campaign is approved & a press release is drafted.

July 16, 1972 The CSPE petition appears on the front page of the Sunday _Contra Costa Times_ with an article by Justin Roberts.

Five days after the petition campaign is announced, the 17 Bay area legislators sign a letter authorizing the Legislative Analyst, A. Alan Post, to conduct an "investigation of the operations of BARTD with particular reference to safety and contract administration."

July 17, 1972 W. A. Bugge, PBTB Project Director, sends a letter to H. J. Walsh, NSPE National Director, questioning CSPE's involvement in the petition campaign. Bugge claims that "It is extremely strange that a national professional society of engineers should allow a relatively small chapter of their organizatioon to take this position without clearance from either the state organization or the national."

July 26, 1972 A joint meeting of the Golden Gate, East Bay & Diablo Chapters of CSPE is held.

August, 1972 Before partial revenue service is started, PUC requires BART to adopt a "Manual Block system" - a train cannot leave one station until a phone call assures that no train is on the track between the stations.

A. Alan Post claims that tests done during August, 1972, clearly show deficiencies in the train detection circuits. Pre-revenue service tests reveal that the train detection subsystem of the ATC would not consistently detect the presence of unpowered transit cars on the tracks.

The CSPE "Diablo chapter's petition campaign was complete."

The Diablo chapter comes under fire by NSPE's Profesional Engineers in Private Practice division.

First Week in The NSPE attorney (Milton Lunch) meets with Anderson and John Kane and suggests that funds would be available from the NSPE legal fund for the legal support of the 3 BARTD engineers. John Kane, Editor of _Professional Engineer_, requests an article from Roy Anderson for the September issue. The article is finished in October 1972, but it is never published.

August 12, 1972 The CSPE Executive Committee passes a motion (with one dissenting vote) to censure the Diablo chapter. No notice was given to the Diablo chapter of the vote.

August 17, 1972 "Ed Walker" sends a letter to Robert Kuntz claiming that the CSPE Diablo Chapter's petition violates the National Society of Professional Engineers' Code of Ethics. According to Ed Walker, the letter was actually written by Jim Walsh who also signed Walker's name.

August 18, 1972 At the Professional Engineers in Private Practice (PEPP) meeting, Jones recommends support of an action to censure the Diablo Chapter.

August 19, 1972 The CSPE Executive Committee recommends that the Board of Directors censure the Diablo chapter; the subject of censure is deferred.

The CSPE Board of Directors appoints an ad hoc committee (consisting of Glenn Wilson, CSPE Vice President South, Bill Holden, CSPE member, William Tarman, CSPE member, Billie Schmidt, CSPE Vice President North) "to confer with 3 BART representatives on general technical subject matter." According to Wilson, they are never able to meet with BARTD management.

September 1972 Three articles appear in NSPE's Professional Engineer praising BARTD. The articles are written by Bill Bugge, project director for PBTB, Dave Hammond, BARTD Assistant General Manager, and Westinghouse. No mention is made of the 3 fired engineers.

September 1, 1972 This is the date Stokes publicly announces that revenue service will begin.

Semtember 2, 1972 PUC conducts "tests" of the ATC system.

September 5, 1972 A fire door in a subway south of Lake Merritt Station falls from its track against a transit car and causes $5000 in damage. Repairs should be completed by March 1, 1974 (due to a delay in getting parts & the low priority of the job).

September 11, 1972 BART opens the first 28 miles of the system between Fremont & MacArthur Stations (downtown Oakland) for revenue service at 12 Noon (almost 3 years behind schedule). Revenue service starts under a special order imposed by the PUC that all trains be separated by 2 stations using a `Manual Block` procedure.

Verdugo sends a letter to Kuntz protesting the censure move without notification to the Diablo chapter that the censure move was taking place. The letter also requests an ethical practices hearing.

Kuntz appoints an Ad Hoc Committee on Ethical Practice to investigate the methods used in the Diablo chapter petition. (Members are Traver Smith, chairman, Eugene Sullivan, CSPE Santa Clara Valley Chapter President, Howard Grant, and Harold Williamson.)

September 13, 1972 Robert Nichols (NSPE PEPP chairman) sends a letter to Kuntz claiming that "It is very distressing that a chapter within NSPE has seen fit to raise such serious questions concerning an area of practice of engineering."

September 22, 1972 Verdugo talks to Walsh & indicates that he would like to meet with the Golden Gate Chapter to explain the situation with the 3 engineers.

September 27, 1972 President Richard Nixon visits BART and rides a train from San Leandro to Lake Merritt Station.

October 2, 1972 BART's 1st accident - a train plunges off the end of the line at Fremont. Four passengers and the attendant are injured. A Crystal oscillator in the ATC subsystem transmitted the wrong speed command.

October 11, 1972 The BART system is officially dedicated by U.S. Secretary of Transportation John A. Volpe at Lake Merritt Station.

November, 1972 Roy Anderson & Justin Roberts try to find a lawyer for the 3 engineers and Anderson tries to get money from the CSPE legal defense fund for a retainer, but no release can be obtained.

November 9, 1972 The CSPE Ad Hoc Committee on Ethical Practice (Trevor Smith, Eugene Sullivan, Howard Grant, and Harold Williamson) meets in San Francisco to hear charges against the Diablo chapter. CSPE Past President Jones presents the charges against the Diablo chapter. Anderson & Verdugo defend the chapter. A CSPE committee commends the Diablo chapter.

The Legislative Analyst's (first) report (also called "The Post Report") is released. The report is critical of the operation and administration of BARTD. The investigation started by examining the June 19, 1972 CSPE report on BART (a report "initially submitted by the Walnut Creek chapter of CSPE"). The Legislative Analyst's report consists of 106 pages of comments & 31 recommendations alledging BART is unsafe as it is being operated.

B. R. Stokes, BARTD General Manager, and George Silliman, President of the BARTD Board of Directors, respond that the problems noted in the Post report are only minor & will be resolved well in advance of starting Transbay service on September 24, 1973. Westinghouse and Rohr Corp. claim no problems exist.

November 14, 1972 The California Senate Public Utilities and Corporations Committee holds its first hearing on BART.

November 16, 1972 Willard Wattenburg, a consultant, sends an open letter to the BARTD Board of Directors & the California PUC criticizing the ATC system.

November 17, 1972 The CSPE Ad Hoc Committee on Ethical Practice report (see September 11, 1972 and November 9, 1972) claims that the Diablo Chapter has not violated the NSPE Code of Ethics.

November 18, 1972 The CSPE Ad Hoc Committee on Ethical Practice report is presented to the CSPE Board of Directors & is adopted.

November 20, 1972 Richard Gravelle (Assistant Chief Counsel) asks Blankenzee to testify at a Public Utilities Commission hearing on the 10/2/72 BART accident.

The BARTD Board of Directors adopts a resolution stating that the system is operating safely by a vote of 10 to 1.

November 21, 1972 The California Senate Public Utilities and Corporations Committee holds its second hearing on BART.

BARTD files a written response to the Post report of November 9, 1972.

November 24, 1972 A $151 million taxpayers' suit is filed in the Superior Court of San Francisco against PBTB.

November 27, 1972 A memo to "Bechtel Senior Management" from the San Francisco office of Bechtel, Inc. claims that "events leading up to ... California Senate PUC Committee hearing - convened earlier that month - had their origins in 1971 (or earlier), when several BARTD engineering employees were dismissed after publicly objecting to certain BART policies."

The California PUC conducts a 2-day field inspection of BART train operations; 100 instances of equipment failures are noted.

November 28, 1972 Kuntz sends the Ad Hoc Committee report to NSPE President James F. Shivler.

The California Senate Public Utilities and Corporations Committee holds its third hearing on BART.

November 29, 1972 Blankenzee sends a letter to Gravelle (PUC Assistant Chief Counsel) with his testimony to the California Public Utilities Commission investigating the 10/2/72 accident. (See November 20, 1972)

Late November 1972 Roy Anderson leaves California for a job with the Transportation Safety Board in Washington, D.C.

December, 1972 Roy Anderson calls John Kane, Editor of <u>Professional Engineer</u>, to inquire about the BART article he submitted in August 1972. Kane says the subject of the 3 engineers & the Diablo chapter is "too hot" at the national level of NSPE & can not be published.

Justin Roberts 9-month series of articles on BARTD in the <u>Contra Costa Times</u>, a Walnut Creek newspaper, ends in December, 1972.

At a press conference, Woodrow Johnson, Westinghouse Vice President, says the train control system is "as safe or safer than any other designed & operating anywhere in the country." He claims the controversy is caused by "an honest difference of opinion among technical people."

December 12, 1972 The millionth rider is carried on BART under revenue service.

December 19, 1972 A special 3-man "Blue-Ribbon Panel" (composed of Drs. Bernard Oliver of Hewlett-Packard, Clarence Lovell, an independent consultant, and William Brobeck, President of William M. Brobeck Assoc.) is appointed by the California Senate Public Utilities & Corporations Committee. The Committee begins its study of alledged deficiencies in the BART system.

December 20, 1972 The final section of rail is fastened down, completing the present BART system of 160 miles of mainline & yard trackage.

January, 1973 The Diablo chapter of CSPE submits Justin Roberts' name for the NSPE Journalism Award.

January 2, 1973 Billie Schmidt, Jim Walsh, Gil Verdugo, Bob Hutchinson, Dick Klouse, Don Leadbetter, and Ed Walker meet and discuss the concerns of NSPE about CSPE affiliation with the Western Council of Engineers & BARTD. Verdugo says their goals are: 1) punitive damages, 2) back pay, and 3) to clear the names of the 3 engineers.

January 4, 1973 Contra Costa Mayors' Conference - Helix talks about BART & says that the 3 engineers approached him & other directors in January, 1972, expressing concern about the reliability of the ATC system & the need to involve lower-echelon staff engineers in testing. Helix claims the engineers hired an outside consultant.

January 29, 1973 The Richmond-Berkeley BART line is opened, adding 11 miles to the system.

January 31, 1973 The Blue-Ribbon Panel (see December 19, 1972) appointed by the California Senate Public Utilities and Corporations Committee reports 21 recommendations for the BART system with five major design modifications and back-ups in the ATC system in order to provide for passenger safety.

Late January, 1973 BARTD Assistant General Manager David Hammond submits his resignation effective March 1, 1973.

February, 1973 The first mention of the BARTD engineers in NSPE's <u>Professional Engineer</u> occurs. It is the only mention of the BARTD case except for the June 1973 journalism award won by Justin Roberts.

February 1, 1973 Ed Walker, Dave Hammond, Bill Bugge, Billy Schmidt, Jim Walsh, and Don Leadbetter meet to discuss BARTD's position.

February 5, 1973 Hjortsvang sends a letter to Helix asking him to create a new position for him in BARTD to fulfill the Post report's recommendations. (See November 9, 1972)

February 17, 1973 Gil Verdugo is appointed Chairman of the CSPE Transportation Safetv Committee.

March 1, 1973 Assistant General Manager ("Chief Engineer") of BART David Hammond resigns.

March, 1973 BARTD management tells the California Senate Public Utilities and Corporations Committee that some of the modifications recommended by the special panel would be implemented prior to starting full revenue service on September 24, 1973. The Senate committee asks the special panel & the Legislative Analyst's office to continue monitoring BARTD's implementation of the recommended modifications.

March 27, 1973 BARTD submits a formal report to the California Senate Public Utilities & Corps. Committee.

May 21, 1973 Hjortsvang, Blankenzee, and Bruder file suit against BARTD for breach of contract and deprivation of constitutional rights. They ask for a total of $875,000 in damages.

Steve Unger, Columbia University professor, sends an open letter of invitation to IEEE members to join a working group on ethics. He says he is gathering information on the BARTD case.

The Concord BART line opens adding 19 miles to the system.

May/June, 1973 Ed Walker's letter to the Editor in support of BART and its consultants appears in the _California Professional Engineer_.

June, 1973 BARTD organizes a start-up task force and begins to reorganize.

Wattenburg meets with Stokes & other representatives of BARTD. He is unable to obtain any ATC test results.

Justin Roberts is awarded an NSPE journalism prize by a committee of engineering & journalism school deans.

June 17, 1973 Wattenburg calls BARTD "Watergate on Wheels" in a letter to the BARTD Board of Directors.

June 25, 1973 BARTD submits a formal report to the California Senate Public Utilities & Corps. Committee.

June 27, 1973 IEEE's Committee on Social Implications of Technology (CSIT) instructs Steve Unger to continue his investigation of the BARTD case and to bring recommendations to the Committee for possible action. (See May 21, 1973)

July 2, 1973 An employee strike stops BART service until August 6.

August, 1973 Sixty of BART's 75 total miles are in revenue service.

August 6-7, 1973 Two incidents of injuries resulting from a gap between the platform & the floor of the transit cars occur.

August 8, 1973 The National Transportation Safety Board report on the October 2, 1972 BART accident is released. It recommends the abandonment of the fail-safe concept & that BARTD management initiate their own independent safety analysis to identify hazards & determine alternate methods to remove the hazards & reduce the risks of full operation.

August 10, 1973 The first BART train travels through the Transbay Tube to the Montgomery Street Station in San Francisco, averaging 70 MPH westward and 80 MPH eastward.

August 21, 1973 A woman falls in a BART train & files an injury claim against BARTD.

September 11, 1973 This is the first anniversary of BART revenue service, with 56 miles in operation and 5 million passengers carried.

September 24, 1973 BARTD assures the California Senate Public Utilities and Corporations Committee that they will implement the "Blue Ribbon Panel" recommendations before starting revenue service under fully automatic controls which was scheduled to start today. (See January 31, 1973)

The publicly announced opening date of the Transbay Tube is not met. BART is unable to start Transbay service because of unresolved problems including an inadequate number of transit cars.

September 28, 1973 The Arthur D. Little report, requested by the BARTD Board of Directors, on organization & staffing is issued. BARTD announces that the position of Assistant General Manager for Operations & Engineering will be reestablished.

October 10, 1973 This is the self-imposed deadline for Rohr to have the first 250 BART cars delivered. (It is 16 months late.)

October 11, 1973 The BARTD Board of Directors meets. The members receive a copy of the Arthur D. Little report outlining how BARTD could save $1.9 to $2.2 million annually. (See September 28, 1973)

October 15-16, 1973 None of 21 recommendations of the "Blue Ribbon Panel" (see January 31, 1973) is as yet fully implemented.

October 23, 1973 100 million passenger miles have been travelled by the BART system since September 11, 1972.

The Legislative Analyst's Second Report is published. It continues to question BART's safety and management.

The California Senate Public Utilities and Corporations Committee holds a hearing on the safety of BART.

BARTD submits a formal report to the California Senate Public Utilities and Corporations Committee.

October 24, 1973 The California Senate Public Utilities and Corporations Committee holds a hearing on the "fiscal condition of BART and increase PUC jurisdiction."

November, 1973 The Urban Mass Transit Administration announces $34 million in additional funding for BARTD.

BARTD hires TRW to perform a 1-year, $500,000 study to resolve its reliability & train availability problems.

November 1, 1973 This is the starting date for the Director of Systems Engineering who will assume full responsibility for the ATC system.

November 3, 1973 The San Francisco BARTD Festival celebrates the opening of the 7.5 mile San Francisco line serving 8 stations.

November 10, 1973 IEEE's Committee on Social Implications of Technology (CSIT) meets in New York City and hears Steve Unger's report on the BARTD case. (See June 27, 1973) He notes the need for a new IEEE mechanism to support the ethical engineer. The Committee asks Unger to obtain information on how to establish a legal defense fund.

December 12, 1973 A backup system to the ATC is recommended by the Senate special panel, Westinghouse, Stewart-Warner, Hewlett-Packard, California PUC, PBTB, TRW, Lawrence Berkeley Laboratory, BARTD, and Legislative Analyst's Office representatives.

BARTD Board of Directors selects Westinghouse to implement the backup system for train detection.

BARTD General Manager Stokes assures the Board of Directors that the $1.3 million payment to Westinghouse will be recoverable because sufficient safeguards had been established to protect BARTD's legal position relative to Westinghouse.

December 26, 1973 California State Senator Alfred Alquist writes a letter requesting the help of the Lawrence Berkeley Laboratory to study BART. (See January 1974)

December 31, 1973 So far, BARTD has paid PBTB $140,507,163 for its services; the additional payment projection is $2,534,000.

January, 1974 The California Senate Public Utilities and Corporations Committee, with the Legislative Analyst's recommendation, retains the Lawrence Berkeley Laboratory of the University of California to evaluate the corrective strategies used by BARTD & Westinghouse on the ATC system.

The California State Senate Committee on Public Utilities & Corporations determines that BARTD will begin to incur a deficit during the 1974-5 fiscal year.

January 15, 1974 The Lawrence Berkeley Laboratory begins to investigate the ATC system.

January 23, 1974 BARTD refuses to accept delivery of any more cars from Rohr Corp. until the seats are properly installed in the cars already delivered.

February, 1974 As of February, 1974, Rohr had delivered 279 transit cars out of a total of 450 on order; of the 279, only an average of 137 were available for service daily.

February 14, 1974 The BARTD Board of Directors decides that an embargo against delivery of additional transit vehicles and payment to Rohr Corp. is to continue until lifted by the Board pending review of Rohr's performance under its contract with BARTD.

February 26, 1974 The California Senate Public Utilities and Corporations Committee holds the first hearing on the economic and fiscal problems of BART.

February 28, 1974 None of the ATC system & only 49 transit cars have reached final acceptance.

March, 1974 TRW, Inc. (hired by BARTD) reports that the ATC system is adequate and that a back-up system is not necessary.

The Legislative Analyst's report claims that BARTD's technical problems are the result of inadequate contract specifications provided by PBTB, and recommends BARTD file suit against PBTB. It also recommends the replacement of B. R. Stokes.

BARTD's projected financial deficit is approaching $100 million over the next 5-year period, according to the Legislative Analyst's report.

March 5, 1974 The California Senate Public Utilities and Corporations Committee holds the second hearing on the economic and fiscal problems of BART.

March 11-15, 1974 The eight San Francisco BART stations are closed by BART management due to picketing by San Francisco municipal employees involved in a citywide strike.

March 12, 1974 The California Senate Public Utilities and Corporations Committee holds the third hearing on the economic and fiscal problems of BART.

March 19, 1974 The third Legislative Analyst's report is issued.

The California Senate Public Utilities and Corporations Committee holds the fourth hearing on the economic and fiscal problems of BART.

March 25, 1974 IEEE's CSIT holds an open meeting in New York City. A resolution calling for IEEE support for "engineers whose acts in conformity with ethical principles may thus have placed them in jeopardy" and specifically for the three BARTD engineers is passed unanimously and sent to the Technical Activities Board (TAB). (See May 17, 1974)

April 23, 1974 The California Senate Public Utilities and Corportions Committee holds the fifth hearing on the economic and fiscal problems of BART.

May, 1974 Dr. Bernard Oliver accuses Post of misstating the facts of Oliver's recommendations and the reliability of the ATC system.

May 1, 1974 The adjusted opening date of the Transbay Tube is not met.

May 17, 1974 The CSIT resolution is considered by TAB, passed and sent to the IEEE Board of Directors.

May 18-19, 1974 The CSIT resolution is received and discussed by the Executive Committee of the IEEE Board of Directors. The Committee instructs IEEE President John Guarrera to form a task force to recommend a policy to be followed by IEEE in support of ethical engineers. Guarrera appoints Bob Saunders task force chairman and Frank Barnes, Hal Goldberg, Bob Shuffler, and Leo Young, members. Guarrera also refers the CSIT resolution to the United States Activities Committee (USAC) Ethics and Employment Practices Committee.

May 21, 1974 The California Senate Public Utilities and Corporations Committee holds the sixth hearing on the economic and fiscal problems of BART.

May 23, 1974 Roy Anderson submits an article on BARTD to Gordon Friedlander. The article is never published but Friedlander uses it as a reference for his article on the 3 engineers in IEEE Spectrum.

June 3, 1974 "New" projected day for opening of Transbay service.

June 4, 1974 Citizens of the BART District vote to make the BARTD Board of Directors elective.

IEEE's CSIT Newsletter publishes an article extremely critical of "CSPE's adverse treatment of Diablo chapter and foot dragging for support of the 3 engineers". Roy Anderson believes this article is responsible for the CSPE move to file an amicus curiae brief in the case of the 3 engineers vs. BARTD.

June 30, 1974 Governor Ronald Reagan signs AB3043 into law, establishing the voting districts from which a 9-member BARTD Board of Directors would be elected in November.

July 1, 1974 General Manager B.R. Stokes resigns after 17 years with BARTD (11 as General Manager). He is succeeded as acting General Manager by Lawrence D. Dahms (former Assistant General Manager for Operations).

Stokes becomes the Executive Director of the American Public Transit Association in Washington, D.C.

September, 1974 CSPE President Leendert Schonewille appoints an ad hoc committee which initiates filing of an amicus curiae brief on behalf of the 3 engineers; the brief is never filed.

September 1, 1974 This date is targeted as the opening date of the Transbay Tube in March, 1974.

September 8, 1974 The Ethics subcommittee of the USAC Committee on Ethics and Employment Practices recommends that IEEE support the 3 BARTD engineers. Questions concerning the ethical behavior of the 3 engineers are raised, and the issue is sent back to the subcommittee for further study.

September 12, 1974 The CSPE "Ad Hoc BART Brief Committee" (Harry Lalor, John Haff, Gil Verdugo, Bruce Anthony, Harold Williamson, B. A. Schmidt, & Ed Walker) is appointed by a letter from Schonewille.

September 16, 1974 Transbay Tube service begins.

The CSPE Ad Hoc BARTD Brief Committee meets and decides to file an amicus curiae brief in the case of the 3 engineers vs. BARTD.

September 26, 1974 A California Senate bill extends the 1/2 cent BARTD sales tax for 2 years to provide $82.2 million as an interim operating subsidy. The bill (SB 1966) had been introduced by Senator James Mills (D. San Diego).

October, 1974 The BARTD Board of Directors votes 7 to 5 to file suit against PBTB, Rohr, Bulova, & Westinghouse. (See November 18, 1974)

October 4, 1974 BARTD moves that the engineers' trial be postponed from October 21, 1974 to at least 60 days later.

October 21, 1974 Original date for the engineers' lawsuit to go to court.

October 26, 1974 Joel Snyder, Frank Cummings, Mike Sassano, and Victor Zourides meet at the Holiday Inn at LaGuardia Airport in New York City. The idea for an amicus curiae brief based on defining a proper standard of ethical behavior for employed engineers is developed. Snyder & Cummings fly to Washington to present the idea to the USAC Steering Committee meeting. The idea is accepted and a draft is ordered prepared for the USAC and Board of Directors meetings to be held in December, 1974.

November 2-3, 1974 The IEEE Executive Committee meets and Stern (IEEE President-Elect) moves that IEEE file a friend of the court brief in the 3 engineers' case against BARTD. The motion is unanimously approved.

November 5, 1974 A new 9-member BARTD Board of Directors is elected to replace the previous 12-member appointed board.

November 18, 1974 BARTD files a $237.8 million suit against PBTB, Westinghouse, Rohr, & Bulova Watch Co. for breach of contract & warranty.

November 26, 1974 Harry Lalor tells the CSPE Ad Hoc BARTD Brief Committee that Attorney Charles Luckhardt doubts that CSPE would be allowed to file an amicus curiae brief in the case of the 3 engineers vs. BARTD. Nevertheless, the committee members are asked to suggest what the brief should say.

December, 1974 A private newsletter reports that IEEE may file an amicus curiae brief over the objections of IEEE Vice President J. K. Dillard of Westinghouse.

Bob Saunders, IEEE Director, reports to the IEEE Board of Directors that the ad hoc policy committee had met 3 times and recommends that IEEE file an amicus curiae brief in the BARTD case as recommended by USAC. (See October 26, 1974)

December 2, 1974 The first elective BARTD Board of Directors is formally installed.

December 4, 1974 The IEEE Executive Committee approves the following 2 recommendations: (1) that USAC be instructed to report, for the action by the Executive Committee and Board, procedures to handle reports of non-adherence to the Employment Guidelines, and the Code of Ethics by March 15, 1975; and (2) that the ad hoc committee be discharged. (See December 1974 and May 18-19, 1974)

December 5-6, 1974 The IEEE Board of Directors meets and approves 3 recommendations of Saunder's committee (See December 4, 1974): (1) The Executive Committee is impowered to enter an amicus curiae brief in any case where an engineer is "taking a position on a matter of ethical principle; (2) The Committee may publicize such actions; and (3) IEEE policy will be to not take an adversary position or intervene on behalf of or against any member involved in a matter of ethical principle. The Board also approves a new IEEE Code of Ethics.

following December 5 Frank Cummings, USAC legal counsel, drafts an amicus curiae brief with the help of Joel Snyder. A 1912 Code of Ethics adopted by the American Institute of Electrical Engineers (one of the original societies of the IEEE) is discovered and incorporated into the brief.

December 18, 1974 A management audit of BARTD is completed by Cresap, McCormick, and Paget.
December 20, 1974 This is the deadline Harry Lalor set for CSPE's Ad Hoc BARTD Brief Committee members to suggest to him what the CSPE brief should say.
January, 1975 Three on-track vehicle accidents occur at BART.
January 4, 1975 CSPE is authorized by the court to file a brief in the engineers' case; CSPE 1st Vice President (Lalor) asks the Secretary to solicit more funds in the Calfornia Professional Engineer magazine.
January 9, 1975 IEEE files an amicus curiae brief in the case of "Hjortsvang vs. San Francisco Bay Area Rapid Transit District and ten Does."
January 16, 1975 BARTD files a motion asking the court to order the 3 engineers to a hearing before a special committee of the BARTD Board of Directors to decide if there was cause for their dismissal.
January 19, 1975 A workman is killed when a 70 MPH test train hits a rail maintenance vehicle causing $700,000 in damage. The ATC system could not detect the maintenance vehicle because it had rubber wheels.
January 22, 1975 State Senator Alfred Alquist asks Lawrence Berkeley Laboratory to investigate the cause of the January 19th BART accident and make safety recommendations.
January 29, 1975 The 3 engineers' lawsuit against BARTD is settled out of court for $25,000 each less 40% for lawyer's fees. The suit was reportedly "flawed" because the engineers had denied hiring an outside consulting firm when asked by management if they had done so.
February, 1975 The new IEEE Code of Ethics is published in Spectrum. (See December 5-6, 1974)
February 3, 1975 The 3 engineers' lawsuit against BARTD was originally scheduled to be heard by Judge George W. Phillips, Jr. in the Alameda County Superior Court in Oakland, CA on this date.
February 4, 1975 The California Public Utilities Commission orders an investigation into "BART's safety appliances and procedures."
February 5, 1975 Harry Lalor sends a copy of the IEEE amicus curiae brief to Robert Kuntz. Charles Luckhardt suggested that CSPE participate in the brief. The case has already been settled out-of-court, however.
February 6, 1975 A Board of Inquiry finds the January 19 BART accident a result of human error.
February 7, 1975 A task force comprised of BARTD employees, Lawrence Berkeley Laboratory, PUC, & PUC consultants is organized.
February 10, 1975 The BARTD Board of Directors approves the out of court settlement for the 3 engineers.
February 15, 1975 Gil Verdugo reports to the CSPE Board of Directors that the BARTD engineers' case has been settled out of court.
Preliminary results from task force field tests demonstate one method of providing maintenance vehicle detection.
April 1, 1975 A task force is established to address procedures & equipment reliability. A vehicle reliability task force, central computer task force & wayside and station equipment task force are established by the California Public Utilities Commission (PUC).
April 24, 1975 Frank C. Herringer is appointed BARTD General Manager, effective July 1, succeeding Lawrence D. Dahms. Herringer was Administrator of the U.S. Urban Mass Transportation Administration in Washington, D.C.

June, 1975 The California PUC issues an interim order that BARTD revenue and non-revenue operations are not as safe as desirable and feasible.

July 1, 1975 Frank C. Herringer officially becomes BARTD General Manager.

July 8, 1975 BARTD responds to the special panel appointed by the California Senate Public Utilities and Corporations Committee.

August 14, 1975 The BARTD Board of Directors adopts a proposal to increase fares effective November, 1975, and charge for parking effective July, 1976.

July 8, 1975 The California Senate Public Utilities and Corporations Committee holds a hearing on BART.

October 23, 1975 Robert D. Gallaway is appointed Assistant General Manager of Operations, effective November 3. He was Executive Vice President for Operations at Texas International Airlines in Houston.

November 3, 1975 The first fare increase since the system opened in 1972 goes into effect, resulting in an average 21% increase in trip fares.

January 1, 1976 Permanent night service goes into effect, extending BART revenue service hours from 6 a.m. to 12 midnight.

January 30, 1976 BARTD is selected as one of the 200 examples of outstanding community achievement in the U.S. as part of the American Revolution Bicentennial Administration's "Horizons on Display" program.

May 27, 1976 The Embarcadero Station is officially opened for revenue service.

July 8, 1976 BARTD management and labor officials sign a new three-year collective bargaining contract after extensive negotiations.

October 15, 1976 Oakland officially dedicates its new City Center Plaza, the Harold Paris sculpture and BART's entrance to the plaza.

November 1976 BARTD has received $315 million in federal capital grants up to this date. Federal aid is 20% of the total $1.6 billion investment in the system.

July, 1977 BARTD wins a $28 million out-of-court settlement against Westinghouse, Rohr, and PBTB. This is approximately 12% of the original claim filed in November, 1974.

Bibliography

Ewing, David. 1977a. Freedom Inside the Organization: Bringing Civil Liberties to the Workplace. New York: E. P. Dutton.

------------ 1977b. "What Business Thinks About Employee Rights." Harvard Business Review, (September-October): 81-93

Glaser, Barney G. 1964. Organizational Scientists. Indianapolis: Bobbs-Merrill.

Goldstein, Sanford. 1977. "Rashomon, Co-Causality, and Me." The Western Humanities Review, 31(Spring): 157-64.

Kornhauser, William. 1963. Scientists in Industry. Berkeley: University of California Press.

Marcson, Simon. 1960. The Scientist in American Industry. New York: Harper and Row.

Olson, J. 1972. "Engineering Attitudes Toward Professionalism, Employment, and Social Responsibility." Professional Engineer, 42 (No. 8): 30-32.

Perrucci, Robert and Joel Gerstl. 1968. Profession Without Community: Engineers in American Society. New York: Random House.

Ritti, R. Richard. 1971. The Engineer in the Industrial Corporation. New York: Columbia University Press.

Scott, W. Richard. 1966. "Professionals in Bureaucracies: Areas of Conflict." in H. M. Vollmer and D. Mills (eds.) Professionalization. Englewood Cliffs, N.J.: Prentice-Hall.

Stone, Christopher. 1975. Where the Law Ends: The Social Control of Corporate Behavior. New York: Harper and Row.

Weinstein, Deena. 1977. "Bureaucratic Opposition: The Challenge to Authoritarian Abuses at the Workplace." Canadian Journal of Political and Social Theory. 1 (Spring-Summer) 31-46.

Zald, Mayer N. and Michael A. Berger. 1978. "Social Movements in Organizations: Coup d'Etat, Insurgency, and Mass Movements." American Journal of Sociology, 83 (January): 823-61.

About the Authors

Robert M. Anderson, holder of bachelor's, master's and doctoral degrees in electrical engineering and physics from the University of Michigan, is manager of engineering education and training at the General Electric Company. Formerly he was professor of electrical engineering at Purdue University. Author of many articles and technical reports, he has done research in the area of electrical and optical properties of thin semiconducting films and regularly taught a variety of courses concerned with solid-state materials and devices, electro-mechanic field theory, electronics, and the legal, social, and ethical aspects of engineering.

Robert Perrucci, professor of sociology and head of Purdue's Department of Sociology and Anthropology, was educated at the State University of New York and Purdue University. Much of his research concerns the interplay between technology and society, with special emphasis on the study of professions, large-scale organizations, and occupational mobility. He has published over forty research articles and book chapters and is author or coauthor of six books.

Dan E. Schendel, a member of the faculty of Purdue's Krannert Graduate School of Management, teaches marketing, strategic management, and strategic planning courses. The holder of degrees from the University of Wisconsin, Ohio State University, and Stanford University's Graduate School of Business, Professor Schendel is the author of numerous articles and the coauthor of two new books in the area of strategy formulation and strategic management. He has consulted widely for various corporations.

Leon E. Trachtman, educated at Hamilton College and The Johns Hopkins University, is professor of communication and associate dean of the School of Humanities, Social Science, and Education at Purdue. His principal research interests are communication in the scientific community and the social and ethical impact of contemporary science and technology. He has published articles in these fields and is a major contributor to a series of elementary social studies texts. He serves as general editor of Science and Society: A Purdue University Series in Science, Technology, and Human Values.